FRACTAL TIME

Why a Watched Kettle Never Boils

STUDIES OF NONLINEAR PHENOMENA IN LIFE SCIENCE*

Editor-in-Charge: Bruce J. West

*For the complete list of titles in this series, please go to
http://www.worldscibooks.com/series/snpls_series

Studies of Nonlinear Phenomena in Life Science – Vol. 14

FRACTAL TIME
Why a Watched Kettle Never Boils

Susie Vrobel

Institute for Fractal Research, Germany

World Scientific

NEW JERSEY • LONDON • SINGAPORE • BEIJING • SHANGHAI • HONG KONG • TAIPEI • CHENNAI

Published by

World Scientific Publishing Co. Pte. Ltd.

5 Toh Tuck Link, Singapore 596224

USA office: 27 Warren Street, Suite 401-402, Hackensack, NJ 07601

UK office: 57 Shelton Street, Covent Garden, London WC2H 9HE

British Library Cataloguing-in-Publication Data
A catalogue record for this book is available from the British Library.

Studies of Nonlinear Phenomena in Life Science - Vol. 14
FRACTAL TIME
Why a Watched Kettle Never Boils

ISBN-13 978-981-4295-97-0
ISBN-10 981-4295-97-3

Printed in Singapore.

To my parents

Preface

This book tackles the notion of duration and the structure of our Now using a phenomenological approach. We shall look into a number of temporal distortions, correctives and compensatory actions. The main focus will be on our perception of duration and possible ways of modifying it. A model of fractal time will be introduced, which will differentiate between two mutually exclusive temporal dimensions that manifest themselves as simultaneity and succession: Δt_{depth} and Δt_{length}. The temporal fractal perspective, which results from nesting simultaneous levels of description can be shown to determine our estimation and experience of duration and even lead to a perceived reversal of the causal order of events.

Duration and the Now are primary experiences of time and therefore precede physical and semantic concepts of time. We have to take account of the limitations of our perceptual apparatus and the integrative performances of our brains. These constraints and performances have to be considered *a priori* when we set up or evaluate a model of reality. While many such constraints are not accessible via introspection, distortions which arise as a result of compensations carried out by our brains are a blessing in disguise, as they reveal to us the – usually invisible – structure of our temporal perspective.

Corrective distortions within the Now reveal the internal structure of our interface with the rest of the world: a nested, temporal fractal, whose structure changes with every boundary shift we perform. Boundary shifts occur when we change the assignment conditions – that is to say, decide what to assign to the external world and what to the observer-participant. The internalization of the exterior and, *vice versa*, the externalization of the interior can have bizarre consequences: the individual concerned may, for example, incorporate a fake hand or disown a part of his body. When spatial or temporal boundaries become transparent – that is to say, no longer visible from the endo-perspective – a new systemic whole will emerge, which has truly assimilated the external structure. This structure could be a rhythm to which we entrain or a delay which we compensate by means of compensatory anticipation.

The art of nesting and de-nesting – that is to say, of contextualization and de-contextualization – will be explored by looking at how we focus against the background of different emotion- or attention-driven arousal, entrainment and predictability.

We shall see that temporal dilation correlates with such diverse phenomena as stressful situations, synchronization, the compensation of delays (compensatory anticipation), lack of perceived contrast and reduction in perceived complexity. The common denominator which connects all of these is a lack of Δt_{depth}. Too little simultaneity is counter-balanced by an increase in succession (Δt_{depth}).

Humans appear to show a preference for fractal structures, both spatial and temporal ones. This manifests itself in our aesthetic appreciation of fractals, such as the golden ratio, in the visual arts and architecture. In the temporal dimensions, our fractal predisposition shows itself in the abundance of musical structures which resemble pink noise – that is to say, those which display 1/f behaviour.

The formation of gestalts reduces complexity around us. These may come in the shape of cluster formation, the anticipation of delays and as trust – our most powerful anticipatory faculty and reducer of social complexity. In this connection, we shall look at the role of complexity in dynamical diseases and temporal misfits and the impact of entrainment.

Eventually, we shall address the phenomenon of insight, which is a collapse of Δt_{depth} – that its to say, the instantaneous reduction of a broad range of rhythms to a single one. In a nutshell, duration is portrayed as a kind of friction which results from a lack of Δt_{depth} and the subsequent generation of Δt_{length}: succession. A watched kettle never boils because we narrow our fractal temporal perspective by focussing on one level of description, which means that simultaneity almost ceases to exist and succession is perpetually generated instead.

In the Appendices, the three pioneers of fractal spacetime, Garnet Ord, Laurent Nottale and Mohamed El Naschie, summarize their theories against the background of the current state of the art.

Acknowledgements

Many thanks to my husband Barry Baddock for moral support, helpful hints and brushing up the manuscript. For helpful communication, my thanks go to Otto Rössler, Jerry Chandler, Otto van Nieuwenhuijze, Franz-Günter Winkler, Anthony Moore, Bill Seaman, Gary Boyd, Hellmut Löckenhoff and John Hiller. I also wish to express my gratitude to the three pioneers of fractal space-time, Garnet Ord, Laurent Nottale and Mohamed El Naschie, for providing appendices containing their theories in a condensed and readable form. Many thanks to Sam Rohn for providing the cover illustration (© 2008 Sam Rohn).

Susie Vrobel
Bad Nauheim, 26 July 2010

Contents

Chapter 1

When Time Slows Down:
Subjective Duration

When we queue at the post office or listen to a boring lecture, something happens to the texture of time: it becomes viscous, drags and almost comes to a standstill. Most of us are familiar with this feeling of time slowing down and do not give it a second thought. However, the mechanisms at work in such everyday experiences also govern more dramatic events, when time slows down to the extent that we seem capable of almost superhuman perception. Let us look at an especially dramatic example from a police officer's report:

> During a violent shoot-out I looked over, drawn to the sudden mayhem, and was puzzled to see beer cans slowly floating through the air past my face. What was even more puzzling was that they had the word *Federal* printed on the bottom. They turned out to be the shell casings ejected by the officer who was firing next to me. [1]

The policeman must have thought twice before handing in his report. Seeing beer cans flying in slow motion does not look good in such a document. But, in fact, his account was taken very seriously indeed, for he was by no means the first officer in the force to have reported a distorted perception of time in a stressful situation.

In his book *Into the Kill Zone*, David Klinger describes similar experiences of officers caught in moments of high stress, such as armed confrontations. Many of them reported, like the officer who saw the beer cans, that time had slowed down: "Then the guy reached down towards the bulge in his waistband. At that point, things went into slow motion, and I said to myself, 'If he reaches under the shirt, I'm gonna shoot him'." [2]

Some also described the sensation of "tunnel vision" – that is, the field of vision being reduced to a small blob, as when looking down a tunnel: "Another thing I remember is that when the guy turned and started firing, I got tunnel vision on him." [3]

Another common feature of these confrontations was loss of hearing – or, more exactly, a shutdown of auditory perception.

> I knew the guy was shooting at us because I saw him shooting, but I didn't really hear the rounds going off. The audible start-up and 'BANG!' that usually happens when you pull the trigger wasn't there. (...) At the time, I didn't know my partner fired because I didn't hear his shots. [4]

Tunnel vision, failing to hear pistol shots, reading an airborne shell casing – how can distortions like these be explained? What actually happened to that officer when time slowed down and he thought he saw beer cans floating past?

Before taking up these questions, let's look at the experiences of some other types of people who have reported a narrowed focus and a slowing down of time.

Buddhist monks in meditation, for example. Some have talked of a stretching-out of time. There are cases, indeed, of individuals for whom, when focussing on a single object in meditation, time has come to a standstill.

According to the *Mahabharata*, a Sanskrit epic from ancient India, the senses become so inhibited in this intense contemplative state that the individual does not hear, smell, taste or see, nor does he experience touch. All his ideas revolve around the object of meditation – a literal example of single-mindedness. As a result, sensory stimuli are excluded, the mind ceases to imagine – and time slows down.

Exponents of Japanese martial arts have described a condition they call being "in the zone". This, too, is a state of slowed-down time – one which can often be brought about at will. In Aikido and Taekwondo, for example, the focus is on the use of energy to fend off the opponent or to gain control over him. For practitioners of such forms of combat, to enter "the zone" is to be able to see the opponent's movements in slow motion. Consequently, he can be observed in more detail and the observer can react more quickly. [5]

Professional baseball players, too, have reported the sensation of being "in the zone". For them, the tempo and the action of the game slow down, and they become acutely aware of what is happening. American batter Tom Gwynn says that when he was "in the zone", the pitched ball appeared to come in so slowly that he could actually see the seams on it.

This state of total immersion is also known as "flow", a term used by psychologist Mihály Csíkszentmihaílyi. [6] In this state, a person is fully involved in and focused on what he is doing. The experience of flow is accompanied by a high degree of concentration on a narrow focus and complete absorption in a task, up to the point of merging awareness and action.

Fig. 1.1. Absorbed in play.

It is marked, too, by a distorted sense of time and the idea of clear goals which can be effortlessly attained. For instance, musicians – whether as individuals or as part of a group – may experience flow when playing their instrument. Similarly, athletes may achieve their best performances without feeling any pressure. And children can become totally immersed in play to the extent of being lost in it.

Observing the seams of a flying ball or the writing on a floating shell casing, perceiving the tactics of an opponent in decelerated motion, meditative time-stretching – such intense sensations may not have been shared by many of us. But still, the experience of time slowing down is familiar to everyone.

In movies, for example. We readily accept slow motion on the screen, as in the scene in *The Matrix* which features so-called "Bullet Time": a computer-generated form of extreme slow motion, during which details become visible which are normally imperceptible because they happen too fast. For instance, flying bullets move through the air so slowly that they can be intercepted or evaded.

Computer games use similar effects. For instance, "The Bourne Conspiracy", an espionage video game, simulates a slowing down of time: The player can build up the avatar's [7] adrenaline and then spill it in the game to slow down the action, so that everything around the avatar appears to happen in slow motion. During the few seconds that a slowing down of time is simulated, the player can react at what, to other players and their avatars, may seem like lightning speed, as they do not experience the same temporal distortion. [8]

Most of us know how palpably time can slow down when we are deep into an especially absorbing task or activity: a yoga practice or a meditation session are examples of "positive" time dilation, that is to say, a pleasant experience of time stretching. But we also know how mercilessly time can drag out when we are waiting for an overdue bus or trapped in a slow-moving supermarket queue. When we wait for something specific to happen or when we are bored, the texture of time becomes sticky, loses pace and almost comes to a standstill.

How do these distortions come about? To find an answer to this, let's revisit our witnesses to slow motion time – the monks, law enforcement officers, martial arts exponents and baseballers – and see what they have in common and where they differ.

Clearly, one thing they shared in their varied experiences is that parts of their perceptual system shut down and left open a very narrow focus.

For the meditating monks, cutting out sound, smell, vision, taste and touch left them with nothing but their thoughts, in pursuit of *dhyana*: the "single flow of ideas". For the police officers, much of their perceptive apparatus closed down in a state of stress, leaving them to focus on just one aspect – beer cans, a bulge under a waistband, a flying shell casing.

Fig. 1.2. Bored taxi driver [9]: a case of slowed down time.

The Aikido and Taekwondo combatants deliberately induced time to decelerate by focussing on nothing but their opponent's movements. And for the baseballer "in the zone", focussing on the speeding ball, an unexpected amount of detail suddenly became visible.

But how to explain the slowing down of time which was reported in all of these cases? One commonly held belief is that adrenaline activity is responsible – as seems to be the case with the playstation player who spills a quantity of the hero's virtual adrenaline. It is true that, in circumstances of high pressure – in a serious accident, say, or in armed combat – the sensation of time slowing down is often accompanied by an adrenaline rush. Parts of the body can shut out: the heart rate may fall, sensitivity to pain may be lost, together with peripheral vision and the capacity to register sound. And attention can become geared to one tiny aspect of the world, as in tunnel vision. It may be tempting, then, to see a causal relation between adrenaline rush and the slowing of time.

A stressful, intensely dangerous situation, such as an impending car accident, is often accompanied by large adrenaline spillage and by the experience of slow motion time. We have all heard – and, perhaps, can testify from personal experience – how time seems to slow down in such moments. Events and actions seem to pass more slowly than we are accustomed to. This sensation is very often described cinematically: "everything seemed to slow down, like in a film".

Slow motion film scenes are, of course, the result of high-speed photography. The increased number of frames are run through at the regular rate of frames per second, so that a slowing down of time is induced. To put it another way, the temporal resolution of the photographic images is deliberately distorted.

Could our experience of slow motion time under stress also be explained in terms of temporal resolution – like the increased number of frames photographed and projected in a movie? Neuropsychologist David Eagleman tried to find out whether the slowing-of-time experience could be linked to "neural snapshots clicking faster during a high-adrenaline situation" [10]. He asked test subjects to take part in an experiment in which they had to jump off a 50-metre tower into a net. Each subject was asked to estimate the duration of the fall. In contrast to bystanders, the jumpers invariably overestimated the duration. So a correlation between adrenaline spillage and time dilation was shown to exist in this, as in other stressful situations.

Next, the question as to whether these perceptions were due to high-speed neural snapshots was put to the test. All the jumpers had a small computer strapped to the wrist which displayed numbers moving in succession. They flashed by slightly faster than the maximum speed at which individual digital numbers can normally be distinguished. The subjects were asked to read those numbers during their free-fall. But none of them were able to do so, nor were they better at making out the flashing numbers than the bystanders were. So Eagleman concluded that the jumpers did not perceive that time slowed down as a result of being in a high-adrenaline situation – did not, in Eagleman's words "obtain increased temporal resolution during the fall." [11] Instead, he concluded, "because memories are laid down more richly during a frightening situation, the event seems to have taken longer in retrospect." [12]

Eagleman's findings suggest that temporal resolution is not increased as a result of "snapshots" being taken at a faster rate during stressful situations. To put it another way, we do not seem to perceive a larger number of successive stills when we spill adrenaline. While this would rule out higher temporal resolution as the cause of slow motion time in such situations, Eagleman attributes slow motion time to the fact that we lay down more memories when we are under stress. We shall return to this question in Chapter 7 when we look the impact of global and local perspectives on our perception of duration and show that laying down more memories can be a result of higher temporal resolution.

Fig. 1.3. High-speed photography: Higher temporal resolution allows us to perceive more detail. [13]

Although high-adrenaline situations may correlate with the perception of time slowing down, no causal connection has as yet been established. An argument against such a simple causal relation is the fact that no such adrenaline activity occurs in case of the Buddhist monks who inhibit their senses at will. So this cannot be the fundamental cause of the phenomenon of time distortion. To find a general explanation, it is worth turning aside from physical impacts – such as adrenaline rush – and looking, instead, at how time itself is structured.

In normal circumstances, we can hear, see, smell, taste and touch things simultaneously. To put it another way, there is a high degree of simultaneity in the way we perceive the world around us. If we shut out any of our senses or thoughts, then this will automatically reduce the number of things we perceive at the same time. If we go further, and focus on one thing only – like a Buddhist monk focussing on a mantra – then we create even less simultaneity. Our Now is, so to speak, reduced to a one-dimensional perspective, with no other stimuli at work.

Focussing on a reduced perspective means, to all intents and purposes, that fewer things are happening at the same time. For a police officer, a colleague nearby may be firing a gun which, in other circumstances, would be extremely audible as well as visible. But, for a man under stress, with tunnel vision and his hearing shut down, his Now is a highly limited, focussed experience, with no sound to be heard and nothing else to be seen.

Before we go further, there is a surprising effect we should look at. It concerns the question of how we remember the length of time periods. Retrospective time judgement is generally attributed to memory storage or contextual change. The greater the content stored or embedded into a context, the longer is the time interval judged to have been in retrospect. [14]

In his so-called "Armageddon experiments" [15], psychologist J.H. Wearden sought out to test our perceptions of duration. In the experiments, participants were asked to watch a scene from the film *Armageddon* and to spend the same time in a waiting room. Wearden asked participants to estimate the length of time spent in these activities.

According to his findings, participants felt that time seemed to fly during the scene-watching, while it dragged during the waiting period. However, when they had to make a *retrospective* judgement, the results were the opposite. [16] That is to say, despite feeling that time had flown while they were watching the film scene, participants later judged that period as being longer than the time spent in the waiting room.

We may say that, for the participants, time passed subjectively more quickly while watching the film scene because its visual effects and sounds created a lot of contextual change. That is to say, events were nested into ever-new contexts, so that a high degree of simultaneity was created. In the waiting room, on the other hand, there was little contextual change: participants basically concentrated only on waiting. In their Now, deprived of simultaneity, time slowed down.

Why is this so? Why might we judge the length of an event differently, depending on whether we experience it within our Now or we remember it later? The amount of content and contextual change, i.e., how much is happening and is embedded in our Now, seems to be crucial to our judgement of duration. Wearden suggests that judgements made in retrospect are based on how much information was processed during the actual experience: Not much happened in the waiting room, therefore time passed slowly. But we remember a period which did not contain many events as having passed faster than a less eventful one. On the other hand, time may pass quickly while we are watching an action-packed film, but when we look back on it, we judge it to have been long-lasting.

Far from being limited to isolated events, this phenomenon may be extrapolated to the span of a lifetime. Thomas Mann's novel *The Magic Mountain* contains a beautiful account of how differently we judge the duration of eventful and uneventful years. His protagonist Hans Castorp muses on the topic of time and memory in this way:

> Emptiness and monotony may dilate the moment and the hour and make them 'tedious'; the great and greatest periods of time, though, they shorten and fade away even into nothingness. Conversely, rich and interesting content is capable of shortening and quickening the hour and even the actual day; on a large scale, though, it endows the course of time with breadth, weight and solidity, so that eventful years pass much more slowly than those poor, empty light years which the wind blows before it, and which fly away. So, actually, what we call tedium is, rather, a pathological diversion of time, resulting from monotony: in conditions of uninterrupted uniformity, great periods of time shrivel up in a manner which terrifies the heart to death .... [17]

"Rich and interesting content" may be interpreted as "a lot happening at once", i.e., a high degree of simultaneity. We may also say that "breadth, weight and solidity" result from an increase in simultaneity. This simultaneity may include new during-relations which arise in the act of recollection itself: the old content is embedded in new Nows. [18]

In a nutshell, our perception and memory of duration can be described in terms of the degrees of simultaneity experienced (see Table 1.1). To account for these notions, a theory of time is needed which takes account of the way in which simultaneity arises. In pursuit of this, we shall – in Chapter 2 – take a closer look at the nature and structure of our Now – our only window to ourselves and the rest of the world.

Table 1.1. Duration now and in retrospect.

	Duration perceived in the Now	Duration remembered in retrospect
Eventful moments/years: high simultaneity (many parallel events)	Time is contracted, speeds up, "flies"	Long time span
Uneventful moments/years: low simultaneity (few parallel events)	Time is dilated, slows down, "drags"	Short time span

References

1. Alexis Artwohl: Perceptual and Memory Distortion During Officer-Involved Shootings, in John E. Ott (Ed.), *FBI Law Enforcement Bulletin*, Oct. 2002, Vol 71, No 10, FBI, Pennsylvania Ave., Washington, DC.
2. David Klinger: *Into the Kill Zone*. Jossey-Bass, San Francisco, 2004, p. 68.
3. ibid, p. 60.
4. ibid, p. 60.
5. In esoterically tinted literature, this phenomenon is also referred to as "lentation" or "chronokinesis".
6. Mihály Csíkszentmihalyi: *Flow – The Psychology of Optimal Experience*. HarperPerennial, New York 1990.
7. An avatar is a graphical representation of a computer user in computer games or other virtual realities to allow the him to identify with a hero or villain.
8. *The Bourne Conspiracy* by Sierra Entertainment 2008. Developed for Playstation 3 and Xbox 360.
9. Photograph courtesy of Rachel Crawford

10. The Eagleman Lab (http://neuro.bcm.edu/eagleman/time.html)

11. ibid.

12. ibid.

13. Photograph courtesy of Jeffrey Clement

14. J.H. Wearden: The Wrong Tree: Time Perception and Time Experience in the Elderly, in: Measuring the Mind: Speed, Control and Age. Edited by John Duncan, Louise Phillips and Peter McLeod. Oxford University Press 2005, pp. 137ff.

15. ibid.

16. They were asked to read a novel for ten minutes to create a break between the two experiments.

17. Thomas Mann: *Der Zauberberg* Fischer, Frankfurt, 1984, pp. 110-111. First published in 1924. (my translation)

18. Hans Castorp's experiences can be described in more detail against the background of a fractal concept of time, which will be introduced in Chapter 2. For now, the notion of simultaneity is sufficient to explain the differences in perceived and remembered duration.

Chapter 2

The Fractal Structure of the Now: Time's Length, Depth and Density

Duration is something we loosely associate with the notion of succession. A 2-hour train journey, for instance, may be pictured as a sequence of 120 one-minute intervals, lined up along an imaginary time axis, like beads on a string. This string is a one-dimensional structure: it extends on one level of description (hereafter denoted as LOD) only, namely the level of successive minutes. On another LOD, say, the level of successive hours, the extension of the train journey would amount to 2 hours. This would still be a one-dimensional extension, albeit on a different LOD. Admittedly, such a concept of duration would be sufficient for most purposes, such as comparing the temporal lengths of train journeys. However, when we wish to compare the subjective duration of a train journey, we will find that one dimension – namely, succession – is not enough. As we shall seek to demonstrate in this chapter, simultaneity is also needed. We shall also look at how duration comes into existence in the first place, how the combination of successive and simultaneous events leads to the nested structure of the Now and how we can quantify and compare subjective duration.

So how does duration arise? As one of our primary experiences of time, along with succession, simultaneity and the Now, duration is a prerequisite for any kind of experience. [1] What we normally mean when we say something has duration is that it lasts for a while – that is to say, a meaningful entity, such as a train journey, has a temporal extension. To contemplate this matter, the German phenomenologist Edmund Husserl looked at another experience we are all familiar with: Why is it that when we listen to a tune, we hear not just a succession of isolated, uncorrelated notes, but a song, a symphony – in short, a coherent whole? After all, as Husserl points out, when we hear the second note of a tune, we no longer hear the first one.

> I do not, then, in truth, hear the tune, but only the individual note. The fact that the section of the tune which has been played is objective to me, I owe – one is inclined to say – to recollection. And the fact that I do not, having reached this or that note, presume that that was *all*, I owe to anticipatory expectation (...). [2]

Husserl concluded that when we hear a new note in the present, the preceding note, which still lingers on in our memory, shifts into an ever-more-remote past. This past, however, must still be in some form present in our Now, as we would otherwise not be able to put together a tune, but hear only one note at a time. We may say that the past Nows are nested in our present Now. With every new note, the preceding ones slip deeper and deeper down the nesting cascade. Likewise, we may say that the present Now is nested into all future Nows we anticipate when we assume that the tune is not yet over but that further notes will follow, embedding the present one.

Husserl used the terms *retension, consciousness of the present* and *protension* to refer to memory, the Now and anticipation, respectively. The consciousness of the present, the Now, is the place where memory and anticipation meet: the potential culmination point of all retensions and protensions. It holds past events by locating them in their fixed positions – fixed with respect to before-and-after relations – and represents them, in a modified way, within the present Now. Thus, we create a simultaneity of retension, the consciousness of the present and protension.

Such a Now, which hosts both retensions and protensions, cannot be a point or a mere dividing line between past and future. It must be extended and display a nested structure as depicted in Fig. 2.1, which continually grows with every new Now holding the preceding one which, again, embeds the Now which precedes it, and so on.

By giving the Now extension and a structure, Husserl paved the way for a definition of subjective duration which can be derived only from the notions of nestings generated by succession and simultaneity.

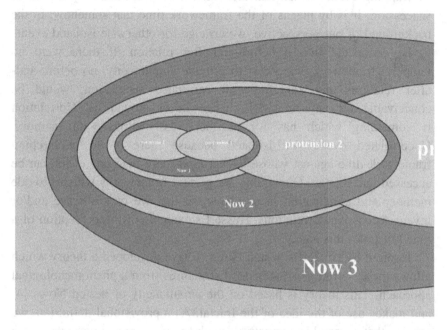

Fig. 2.1. Nested Nows: Now 1 ◈ Now 2 ◈ Now 3 ◈ ... etc. [3]

We may paraphrase succession as before-and-after relations and simultaneity as during-relations. Those during-relations are brought about by embedding an event into a framework time, i.e. by nesting it into a context. This contextualization may occur through remembering and anticipating events, such as the musical notes in Husserl's example. It may also emerge from a nesting of concurrent, partly overlapping actions. During my train journey, for instance, I may be reading a book and drinking a glass of water. In terms of simultaneous relationships, we may say that drinking water ◈ reading a book ◈ sitting in a railway carriage.

We can only make sense of successive events which happen on one LOD if we can arrange them against the background of a context such as a framework time. Individual notes receive their meaning from the entire tune they are embedded in. So, in the wake of Husserl's considerations, a nested structure of the Now can be assumed. Every Now contains overlapping events which add the dimension of simultaneity to that of succession. It is by means of the framework time that somehow, in the background of our perspective, we arrange for otherwise isolated events to be connected into a before-and-after relation. If there were no temporal framework or context to provide simultaneity, no before-and-after relations and, therefore, no correlated succession, would be conceivable for any temporal change in or around us. Correlation is something which has for us loose associations with memory. Uncorrelated events are isolated because there is no overarching framework time against whose background more than one event can be accessed and related. Our brains are perfect frameworks which provide memory and anticipation in our Now, so that we may observe and/or invent successive correlations. Husserl's example of the perception of a tune illustrates this nicely.

Inspired by Husserl's nested Nows, I have developed a theory which allows the description of embedded structures from a phenomenological approach. This theory is based on the simultaneity of nested Nows [4] and makes use of the idea of the fractal. As a provisional definition, we can say that a fractal is a structure which shows more and more detail the higher the resolution. For example, to zoom with Google Earth into an uneven coastline will progressively reveal increasingly jagged detail. We shall soon return to this definition. Let us say for now, though, that the Theory of Fractal Time takes account of all our primary experiences of time (duration, succession, simultaneity and the Now) and allows time series to be quantified and compared. Further, it demystifies the phenomenon of subjective duration.

As Husserl's musical example has shown, one dimension is not enough to describe subjective duration. Mere succession is insufficient – a framework time is necessary, in which a tune can unfold. By adding a second dimension, namely simultaneity, to succession, we are in a position to produce nestings. Based on these two dimensions, the following definitions can be formulated [5, 6, 7]:

- Δt_{length} is the number of incompatible events in a time series, i.e., events which cannot be expressed in terms of during-relations (simultaneity). Δt_{length} defines the temporal dimension of succession for individual LODs.

- Δt_{depth} is the number of compatible events in a time series, i.e., events which can be expressed in terms of during-relations. Δt_{depth} defines the temporal dimension of simultaneity and provides the framework time which allows us to structure events in Δt_{length} on individual LODs.

- $\Delta t_{density}$ is the fractal dimension of a time series [8]. It describes the relation between compatible and incompatible, i.e., successive and simultaneous events – that is to say, the density of time.
  Note that Δt_{depth} logically precedes Δt_{length}, as there is no succession without simultaneity.

Now that the term fractal has reappeared, it is timely to introduce this notion in a more detailed manner, including its sister concept of a fractal dimension. I shall do this with a minimal amount of equations and place more detailed definitions in Appendix A for those readers who wish to explore further. Let us now take a brief excursion into the realm of fractals.

In 1967, Benoît Mandelbrot introduced to the world his notion of the fractal by pondering the question "How long is the coast of Britain?" [9] It turned out that the answer depends on the yardstick you use. If that yardstick is 200 kilometres long, the approximate length of the British coastline will measure 2400 kilometres (see Fig. 2.2). By halving the measuring rod, that length increases to 2800 kilometres (see Fig. 2.3).

If the yardstick is reduced in size still more, say to 50 kilometres, the coastline measures 3400 kilometres (see Fig. 2.4).

The measured length increases because smaller yardsticks take account of more detail on Britain's wrinkled shoreline when they follow the banks of the river deltas and protruding rocks. So as the scale of the measuring rod approaches zero, the measured length of the coastline continues to increase.

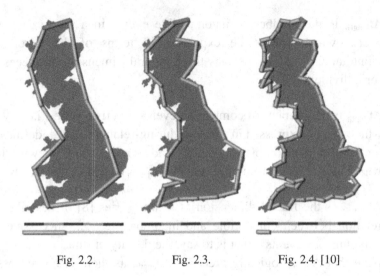

Fig. 2.2.   Fig. 2.3.   Fig. 2.4. [10]

A fractal is most generally defined as a structure which displays increasingly more detail as one zooms into it. The British coastline is an example of such a structure. A feature of fractals is that they are often self-similar. This means that the structure of the whole is found again in its parts, usually in a nested way on several LODs.

Fig. 2.5. Russian Matryoshka dolls. [11]

Nesting alone, however, does not produce fractality or self-similarity. Consider a set of Russian Matryoshka dolls nested inside each other: doll 1 ◈ doll 2 ◈ doll 3 ◈ doll 4 ◈ doll 5. Here, all 5 dolls exist simultaneously as independent wholes (see Fig. 2.5). These do not constitute a self-similar fractal. Self-similarity is the property of a structure in which the whole is actually *made up* of smaller copies of itself. The romanesco cauliflower is an example of this (see Fig. 2.6).

There are different types of self-similarity: perfect, quasi- and statistical self-similarity. The romanesco displays quasi-self-similarity because, although each bud consists of a series of smaller buds, they are not exact copies of the larger structures. Though they may be very similar to the structure of the whole vegetable, the individual buds differ in minuscule ways. An example of statistical self-similarity would be a very meandering coastline, such as that of the British mainland, which, as one zooms in, displays roughly the same intricacy at various resolutions.

Fig. 2.6. Romanesco cauliflower. [12]

Perfect self-similarity is only found in mathematical constructs, i.e. structures which are artificially generated by means of an algorithm. These structures, like the Koch curve depicted in Fig. 2.7, also tend to be infinitely zoomable, i.e. there is no upper boundary, whereas natural fractals exhibit an upper and a lower limit in the range of their self-similarity.

To generate a self-similar fractal, we can use a recursive algorithm. This is a set of instructions which are repeated over and over again. For the Koch curve, these are: take the middle third of a line and replace it with two such line segments, arranged as a protruding spike (see Fig. 2.7, LOD1). Apply this same instruction to all four parts of the new structure (LOD 2) and so on ad infinitum. The length of this curve continues to grow with every iteration by 4/3. To determine its extension, the fractal dimension comes in handy. It measures how densely a structure fills, say, a 2-dimensional space. For self-similar structures, one way of measuring the fractal dimension is the so-called similarity dimension d:

$$d = \log n \, / \log s,$$

where n denotes the number of individual parts on one LOD and s is the scaling factor. So, for the Koch curve, the similarity dimension $d = \log 4 \, / \log 3$, which amounts to approximately 1.2618.

Note that this similarity dimension exceeds the topological dimension of a line, which is 1. To recall, the topological dimension of a point is 0, a line extends in one dimension, a plane is 2-dimensional and a cube measures a topological dimension of 3. With a few notable exceptions, the similarity dimension of fractal structures differs from their topological dimension.

Often people refer to self-similarity as scale-invariance. Although the concepts differ slightly, I shall use these terms interchangeably. The adjectives self-similar, scale-invariant and scaling I also use synonymously.

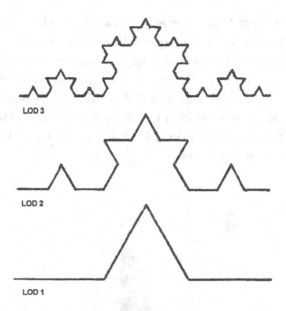

Fig. 2.7. The Koch curve.

Perfectly self-similiar structures such as the Koch curve can be measured with the similarity dimension. So-called quasi-self-similar structures, where zooming, for instance, into a coastline, will produce a similar degree of convolution, but not exact copies of the whole (see Fig. 2.8), cannot be measured with the similarity dimension. The same is true of statistical similarity, which refers to the statistical distribution of a structure over several scales. The reason is that they lack the perfect nested structure, even though they may have been created by a simple recursive algorithm. [13]

To measure quasi- and statistical similarity, we would need to cover the structure with bubbles or boxes which decrease in size as the measurement becomes more precise. Known as the Box Counting Method, this alternative way of measuring the fractal dimension was introduced by Michael Barnsley in 1988. [14] It is useful for determining the extensions of both self-similar and non-self-similar structures. These include structures of a kind that nature tends to produce, whose similarity is such that the parts are not exact replicas of the whole, but only rough copies. Barnsley's method is a generalization since it is applicable both to mathematical and natural structures. Furthermore, it is not limited to "broken" dimensions (such as the Koch-curve's dimension of approximately 1.2618) as even a 2-dimensional plane-filling structure may be described in terms of a fractal by the Box Counting Method. It determines the fractal dimension by calculating d = log n / log s when the scaling factor s approaches ∞.

Fig. 2.8. The Labrador coastline: similar degrees of wrinkling on several scales. [15]

Figures 2.9-2.12 show the coastline of Labrador measured by boxes of length 1/6, 1/12, 1/24 and 1/48. I have skipped the first two steps (length 1 and length 1/3 for the sake of brevity). For a close approach to the fractal dimension, this method needs to be repeated for a very long time, employing smaller and smaller boxes.

Fig. 2.9. Side length 1/6, n = 17, thus d = ln17/ln 6 = 2.833213.../1.791759... = 1.581246 ...

Fig. 2.10. Side length, 1/12 n = 37, therefore d = ln37/ln12 = 3.610917.../2.484906... = 1.453140...

Fig. 2.11. Side length: 1/24 n = 95, therefore d = ln95/ln24 = 4.553876.../3.178053... = 1.432914...

Fig. 2.12. Side length: 1/48 n = 219, therefore d = ln219/ln48 = 5.389071.../3.871201... = 1.392092...

As the scaling factor approaches infinity, the number of boxes approaches a more and more stabilizing value. However, this can be messy during the early stages, where values might fluctuate considerably. For a perfect self-similar structure such as the Koch curve, where there is a recursive generation rule, d remains a stable value throughout the iterations (see Appendix B).

Another remarkable feature of the Box-Counting Method is that it shifts fractality into the eye of the beholder: Even a non-fractal structure can be looked at through "fractal spectacles" consisting of superimposed grids of various scales. The observer determines the yardstick and hence the measurable extension of any structure, be it spatial or temporal. [16]

So far, we have described features of spatial fractals only. However, as we are concerned with temporal nested structures, let us have a look at a few examples of statistically self-similar and perfectly self-similar temporal fractals.

Statistically self-similar temporal intervals can be found, for example, in seismic disturbances. They occur when the distribution of fore- main- and aftershocks is the same over short and long intervals – a phenomenon which has been described by Kagan and Knopoff:

> ... almost all earthquakes are statistically and causally interdependent, a conclusion that contradicts attempts to divide the full catalogue of earthquakes, either into sets of independent or main sequence events (aftershocks and foreshocks). If this picture applies even for the strongest earthquakes, and our results (...) seem to confirm this, then all earthquakes occur in superclusters with very long time spans ... [17]

To illustrate the idea of a perfect self-similar temporal interval, imagine the frequency ratios of musical notes being played simultaneously. An octave is the interval which defines the least complex frequency ratio between two musical notes, i.e. 2:1. For an oboe, for example, the note A has a frequency of 440 Hz. The next higher A would have a frequency of 880 Hz, and so on. These overtones, when sounding simultaneously, generate a nesting cascade which we perceive as consonance. [18] And as the overtones are integer multiples of the fundamental frequency, their structures can be easily translated into each other. We then have an example of a temporal perfectly self-similar fractal structure.

Now if, as in the overtones example, a structure recurs at a different extension in Δt_{length} on nested LODs, it is possible to provide a translation between the individual nested LODs by using that structure as a reference. [19]

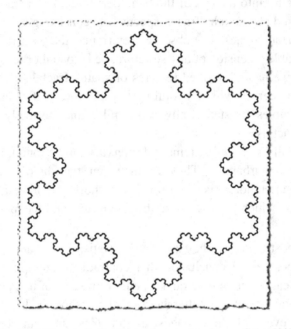

Fig. 2.13. The triadic Koch island or von Koch snowflake. [20]

An ideal example is the triadic Koch island (shown in Fig. 2.13), which can be used as a fractal clock which measures time on an infinite number of nested LODS. Mandelbrot describes the relatively simple generation of this structure:

> The construction begins with an 'initiator', namely a black Δ (equilateral triangle) with sides of unit length. Then one pastes upon the midthird of each side a Δ–shaped peninsula with sides of

1/3. This second stage ends with a star hexagon, or Star of David. The same process of addition of peninsulas is repeated with the Star's sides, and then again and again, ad infinitum. [21]

Δt$_{depth}$: 3 ticks **during** 12 ticks **during** 48 ticks **during** 192 ticks **during** 768 ticks **during** ...

Fig. 2.14. The generation of the triadic Koch island, a.k.a. the von Koch snowflake. [22]

This fractal clock runs with an infinite number of pointers attached to the perimeter of the triadic Koch island, with all pointers ticking away simultaneously, each at its own speed. While pointer 1 ticks only three times (per lap), pointer 2 is ticking twelve times, pointer 3 is ticking 48 times, and so on, ad infinitum (see Fig. 2.14).

Fractal time metric is a generalization of Newtonian metric, which may be defined as a special case of fractal time, as Δt$_{length}$ of the nesting level ∞, with Δt$_{depth}$ = ∞.

The triadic Koch island is, of course, a mathematical idealization which finds no counterpart in the natural world. However, there is an abundance of nested rhythms in nature, including our bodies and minds, albeit for a limited scaling range. Although the spatial and temporal fractals within human physiology and psychology have been discovered only fairly recently in the wake of chaos theory and theories of complexity, the idea of nested rhythm is not an entirely new one.

The oldest example of a clock portraying nested rhythms such as minutes, hours, days and months, is the Prague Orloj, which dates back to the early 15th century (see Fig. 2.15). In addition, it shows Bohemian time (from sunrise to sunset), the lengths of days and nights throughout the year, the moon phases, the ascent and setting of the stars, the signs of the zodiac as well as festive days. Below its clockface is another, which shows seasonal changes and rural life in a sequence of 24 sections.

Fig. 2.15. The Prague Orloj – the first fractal clock.

Why this rather detailed excursion into the notions of fractals and fractal dimensions? As we saw earlier in this chapter, our Now has a nested structure, which can be measured in terms of Δt_{length}, Δt_{depth} and $\Delta t_{density}$. How we perceive duration depends on both our internal degree of complexity (how many nested LODs – that is to say, how much choice of perspectives, do our fractal spectacles provide?) and the degree of complexity our environment offers. Both spatial and temporal fractals are ubiquitous in nature: From the filligree blood vessels in our kidneys to the nested oscillators in our brains – the dynamics which govern our physiology and behaviour display more often than not fractal patterns. Furthermore, the dynamics within our bodies and those of our environment often display statistical self-similarity, i.e., the rate of change or variability of a time series is scale-invariant. There is substantial evidence that a high fractal dimension, which measures the degree of, for instance an individual's heart rate variability, correlates with a healthy state (whereas rigidity, i.e., poor variability, correlates with a lower fractal dimension and disease). [23]

Nested rhythms can be described in terms of fractals if these rhythms actually belong to one systemic whole from which individual parts cannot be separated and described in isolation (as is possible in the example of the Russian dolls presented earlier). An example of nested rhythms within the human body is neural oscillations being embedded in much slower metabolic rhythms. [24]

Outside the body, we find tidal rhythms embedded in astronomical ones. Often, internal rhythms, i.e. those within the body, lock into external ones. This locking-in is also referred to as entrainment. For the human body and mind, entrainment can be both helpful and dangerous. Whether the encounter of internal and external rhythms or oscillations produces a consonance or dissonance is a matter of the individual's ability to encourage or resist entrainment. During meditation, it can be relaxing, for example, if one's brain waves entrain with the external background music. On the other hand, entrainment induced by sound or light has also been known to lead to convulsions. Brief, intense flashes trigger harmonic activity in the brain, i.e. we may say that our brain produces the equivalent to nested overtones – a temporal fractal pattern. Epileptic seizures which are brought on by flashing lights are an example of such photo-convulsive response. [25, 26] Harmonics at three times the stimulation frequency are also produced by other square wave stimulation. This means that people who use meditation stimulation at 10 Hz may find themselves experiencing a panic attack because the 10 Hz stimuli often produce 30 Hz brain waves, which trigger anxiety. [27] Therefore, entrainment is not necessarily a desirable phenomenon and the way to avoid unwanted locking-in of rhythms is to cultivate a large variety of possible responses to external or internal stimuli.

Chronobiology, the study of rhythmic phenomena in living organisms, seeks out desirable and undesirable entrainment – that is to say, the synchronization of internal rhythms with external ones. A well-known example of such entrainment is the wake-sleep rhythm, a so-called circadian rhythm, which is a roughly 24-hour cycle of physiological processes and behavioural patterns. Cycles which are shorter than a day (ultradian) or longer (infradian) are nested into or embed circadian rhythms. This means that within our bodies, the nested or overlapping rhythms form a complex chronobiological pattern, in which different rhythms are coupled. Simple ratios of frequencies such as 4:1 occur frequently, for instance, in the pulse-respiratory quotient during the early morning hours, but also during meditation. [28] When we look at entrainment, where internal rhythms lock into external ones and sometimes produce a fractal rhythm in the shape of harmonics, we take it for granted that our bodies or other living systems can actually produce nested rhythms. However, this requires a fair degree of complexity, which differs from individual to individual. Certain physiological or psychological impacts may increase or reduce my internal complexity: Exposure to coloured light, for instance, influences heart rate variability within minutes. [29] Depending on whether it is increased or lowered, one is more or less capable of responding to both environmental and internal changes. This, just like the unwelcome forms of entrainment mentioned earlier, may also turn an individual into a pathological case – a temporal misfit. More on this topic in Chapter 9, where we shall look at pathological examples resulting from a mismatch of nested rhythms.

Before we move on to the next chapter, which looks into spatial and temporal distortions and describes how our temporal perspectives are formed, I would like briefly to draw attention to an observation by Douglas Hofstadter which is highly relevant to the interpretation of fractal clocks, as these use self-similarity to bridge nested levels. Hofstadter distinguishes between two types of causality: one by translation, where the explanation of a phenomenon is often a description of the same phenomenon on a different LOD, and one by 'normal' physical causality.

Moreover, we will have to admit various types of 'causality': ways in which an event at one level of description can 'cause' events at other levels to happen. Sometimes event A will be said to 'cause' event B simply for the reason that the one is a translation, on another level of description, of the other. [30]

Whether or not this differentiation may play a role in a fractal perspective will be debated in Chapter 4. I have raised the question at this stage because I believe it to be a useful one to keep in mind while exploring the nested world of fractal time.

To conclude, as we saw in Chapter 1, the structure of perceived time is not uniform. It appears to have a texture which can make minutes last forever and years slip by in a flash. This is so because the density of time's texture depends on the number of new contextualizations, i.e. nestings, we carry out. Monotony results in a string of successive moments (Δt_{length}) with little or no simultaneity (Δt_{depth}) being created. We may say – carefully – that the degree of temporal complexity for a bored observer is low. (Carefully, because there are literally dozens of definitions of complexity, as we shall see in the next chapter.) Exciting events, on the other hand, create a high degree of simultaneity (Δt_{depth}) and exhibit a high degree of complexity, as a lot of detail is perceived in parallel.

When we want to describe in both a qualitative and a quantitative manner the internal complexity of the human body, the external complexity of the rest of the world, and the ways we are embedded in our environments, it makes sense to use a nested description. It is this very nesting, both of interal and external structures, which creates simultaneity (Δt_{depth}) and, as we shall see in the next chapter, our temporal perspective – a perspective which is based on the notion of a fractal structure of the Now and a fractal metrics of time series.

The fact that subjective duration arises from the interplay of both internal and external nested structures, is often overlooked by physical theories of space-time which do not take account of our primary experience of time and our embodied perspective. (One may even argue that all physical theories are anthropocentric because they have been developed under the constraints inflicted upon us by the way our bodies and minds have evolved and function. [31]) However, there are exceptions: Apart from my own humble contribution, entire cosmologies which are based on the notion of fractal space-time or scale-relativity have been developed by Garnet Ord, Laurent Nottale, and Mohamed El Naschie. A popular account of Ord's fractal space-time, Nottale's scale-relativity and El Naschie's e-infinity can be found in Appendices C, D and E.

References

1. Cf. Pöppel, E.: Grenzen des Bewußtseins – Wie kommen wir zur Zeit, und wie entsteht die Wirklichkeit? Insel Taschenbuch, Frankfurt 2000. More on this topic in Chapter 6.
2. Husserl, E.: Vorlesungen zur Phänomenologie des inneren Zeitbewußtseins. First published in 1928. Niemeyer 1980, pp. 384 ff (my translation)
3. ❖ denotes "nested"
4. Vrobel, S.: Fraktale Zeit. Draft dissertation 1994, unpublished.
5. Vrobel, S.: Fractal Time. The Institute for Advanced Interdisciplinary Research, Houston 1998.
6. Vrobel, S.: Fractal Time and the Gift of Natural Constraints. Tempos in Science and Nature: Structures, Relations, Complexity. Annals of the New York Academy of Sciences, 1999, Volume 879, pp. 172-179.
7. Cf. Vrobel, S.: Fraktale Zeit. Draft dissertation 1994, unpublished.
8. There are many ways of determining fractal dimensions. We shall have a closer look at Barnsley's box-counting method and the similarity dimension. Others, such as the Hausdorff dimension is dealt with in Appendix A.
9. Technically speaking, he described the notion of a fractal in his 1967 article How Long Is the Coast of Britain? Statistical Self-Similarity and Fractional Dimension. Science, New Series, Vol. 156, No. 3775. (May 5, 1967), pp. 636-638. But he coined the actual term "fractal" much later in his The Fractal Geometry of Nature. W.H. Freeman, San Francisco 1982.
10. Illustrations courtesy of Alexandre Van de Sande, www.wanderingabout.com. Figure 2.2-2.4 may be redistrbuted and modified, provided the creative commons license is acknowledged.
11. Photograph "The Usual Suspects" courtesy of Eskimimi (flickr.com).
12. Photograph courtesy of Tim Moss (flickr.com).
13. An example of a quasi-self-similar structure is the Mandelbrot set, which is generated by a set of simple recursive algorithms.

14. Barnsley, M.: Fractals Everywhere. Academic Press 1988, pp. 176ff.
15. Illustration courtesy of Ned Lyttelton (flickr.com). See also Figs 2.9-2.12.
16. Mandelbrot, B. B.: The Fractal Geometry of Nature. W. H. Freeman, San Francisco 1982.
17. Kagan, Y. Y., L. Knopoff: Stochastic Synthesis of Earthquake Catalogs, in: Journal of Geophysical Research, Vol. 86, No. B4, April 1981, p. 2861.
18. "The idea of consonance is ultimately grounded in the notion of commensurability, an essential in Greek mathematics. We recognise consonance when we perceive a certain number of vibrations of one frequency exactly matching a certain number of another frequency." Fauvel, J. et al: Music and Mathematics – From Pythagoras to Fractals. Oxford University Press 2003, p. 27.
19. This assumes that the observer's internal make-up is complex enough to differentiate between these LODs.
20. Illustration courtesy of Kean Walmsley.
21. Mandelbrot, B. B.: The Fractal Geometry of Nature. W. H. Freeman, San Francisco 1982, p. 42.
22. Illustration courtesy of Kean Walmsley.
23. West, B. J.: Where Medicine Went Wrong – Rediscovering the Path to Complexity. World Scientific, Singapore 2006.
24. Buzsáki, G.: Rhythms of the Brain. Oxford University Press, Oxford, U.K. 2006.
25. Siever, D.: Audio-Visual Entrainment: Safety and Tru-Vu Omniscreen Eyesets. Mind Alive, Edmonton, Canada 2006.
26. Harding; F. A., Jeavons, P. M.: Photosensitive Epilepsy. Lavenham Press, Suffolk, U.K. 1994.
27. Siever, D.: Audio-Visual Entrainment: Safety and Tru-Vu Omniscreen Eyesets. Mind Alive, Edmonton, Canada 2006.
28. Kratky, K. W.: Chronobiology and Cross-Cultural Medicine: Cyclic Processes during a Day, a Year and a Lifetime. IIAS Proceedings of the 24th International Conference on Systems Research, Informatics and Cybernetics. Tecumseh, Canada 2004.
29. Kratky, K. W., Axel Schäfer: The Effect of Colored Illumination in Heart Rate Variability, in: Forschende Komplementärmedizin; 13; 167-173, 2006.
30. Hofstadter, D. R.: Gödel, Escher, Bach: An Eternal Golden Braid. Penguin, London 1980, p. 709.
31. For further discussion of this quesion, see Lakoff, G. & R. E. Núñez Where Mathematics Comes From: How the Embodied Mind Brings Mathematics into Being. Basic Books 2000; and Pfeifer, R. and J. Bongard: How the Body Shapes the Way we Think. MIT Press, Cambridge 2007.

Chapter 3

Fractal Temporal Perspectives:
Corrective Distortions

When we used the terms simultaneity and succession to describe subjective duration, we saw that these temporal extensions are closely linked to the notions of temporal compatibility and incompatibility: Compatible intervals produce Δt_{depth} and incompatible ones generate Δt_{length}. When we perceive distortions, such as the time dilations described in Chapter 1, these tend to be incompatible not only with other people's perspectives, but with our own, "normal", undistorted perception too. In this chapter, we shall look at both spatial and temporal examples of how compatible and incompatible perspectives change our perception of structures or processes. We shall also investigate how correction processes which underlie visual and auditory illusions make it possible to see reference frames we are normally not aware of – that is to say, transparent contextualizations. Here, contextualizing means nesting the past into the present Now and the latter into an anticipated future. Such nesting both expands and constrains the observer's temporal perspective – his Now.

Corrective distortions occur whenever we try – usually unconsciously – to make the world around us appear less complex and more consistent than are our perceptions of it. It is useful, for instance, not to assume that your friend has shrunken dramatically since he started walking down the road. We tend to see a familiar object or person as having a fixed size, shape or colour, even when distance, perspective or colour change. Perceptual constancy arises because our impressions conform to the objects as we assume them to be rather than to the perceptual stimulus. However, this only works if we have had enough experience of watching familiar objects under changing light conditions, or from various

35

perspectives or distances. To have no or little experience of spatial depth, for instance, will lead to misinterpretations. Such was the experience of Kenge, a young Pygmy of about 22, who had grown up in a dense Congo jungle with a visibility of 100 yards at the most. When he was taken out to the plains, he saw a herd of about a hundred buffalo grazing at the horizon. He asked the driver what kind of insects they were and tried to liken them to the beetles he was familiar with. When he was told that these were animals twice as big as the forest buffalo he was accustomed to, he laughed out loud and continued to contemplate what type of insect they were observing. When he was taken closer and finally saw that they were real buffalo, he kept wondering whether they had really been smaller at first and had suddenly grown or whether some form of witchcraft was involved. [1]

This example shows that dealing with spatial depth is an acquired skill. The correction processes which we perform automatically after learning about spatial perspective allow us to identify objects under varying conditions and therefore make life a great deal easier. And although we may not be aware of these conditions, we take them into account when we interpret our perceptions.

A well-known corrective distortion which is based on size constancy occurs in the so-called corridor illusion, a misinterpretation not unlike Kenge's. Most of us experience spatial depth from early childhood on, so we take relative size and distance into account when we look at an object embedded in its surroundings. Because objects remain at the same size *relative to objects surrounding them,* they seem to keep this size, even when we are looking at them from various distances. As with other correction processes, we are usually not aware of this home-made distortion. This changes dramatically, however, when familiar objects are no longer of the same size relative to their surroundings. We then experience the corridor illusion (see Fig. 3.1)

We perceive the two monsters in the corridor as being of different sizes because we take into account their relative positions to the background. The illusion occurs when we interpret a 2-dimensional representation as a 3-dimensional one. If we did not – involuntarily – embed the picture in an additional, albeit physically non-existent, dimension, we would not fall for the illusion. The corridor illusion demonstrates our tendency to automatically contextualize. Although this tendency may appear to be a disadvantage here, it enables us to navigate the world smoothly in everyday life and not mistake buffalos for beetles.

Fig. 3.1. The corridor illusion. [2]

Our ability to embed percepts into physically non-existent structures, and thereby create depth, is not limited to the visual realm. We are just as biased towards creating depth in the acoustic world. When we listen to a sound which is made up of self-similar frequencies and, say, the two lowest are artificially removed, we still hear them even if they never sounded. This phenomenon occurs in overtones and, in musicology, is known as *the missing fundamental*. In Chapter 2, we briefly looked at overtones as examples of perfect temporal self-similarity. Our perception of such acoustic structures with the fundamental frequency removed triggers a completion process, in which we create what is physically non-existent: although only the overtones are played, this does not change the perception of the pitch. Even if one takes away not only the fundamental frequency but also the first overtone (plus the second, third, etc.) this does not change the perception of pitch. This phenomenon can be used to trigger the perception of low frequencies which are physically non-existent – for example, to generate in stereo speakers the low-frequency bass sound which the speakers cannot physically produce. Likewise, in telecommunications, only the higher frequencies are transmitted, as the listener can hear the missing fundamental and the first few overtones, even though they are physically not present in the signal.

Just as in the corridor illusion, our brains make a corrective distortion which creates a context that exists only mentally. The third dimension in the monster picture is no more physically existent than is the fundamental frequency in the acoustic structures we have discussed. Yet we make them up to provide a context we apparently need to perceive a familiar world. This completion process may be the result of the listener embedding himself into a wider anticipated context. It seems to be irrelevant that this wider context is not physically present but created by the listener's expectation – our anticipatory faculty which Husserl referred to as protension [3].

The above example of the missing fundamental is a case of shorter nested intervals of time – that is to say, higher frequencies, triggering the creation of longer intervals, i.e., longer, embedding frequencies. But just as shorter, nested rhythms can trigger the creation of embedding ones, longer rhythms can influence shorter, embedded ones. György Buzsáki observed in the power density of electroencephalograms that shorter, nested temporal intervals – that is to say, higher frequencies, are influenced by the longer temporal intervals, i.e. by lower frequencies, which embed them.

(...) This $1/f^{\alpha}$ power relationship implies that perturbations occuring at slow frequencies can cause a cascade of energy dissipation at higher frequencies, with the consequence that widespread slow oscillations modulate faster local events. The scale freedom, represented by the $1/f^{\alpha}$ statistics, is a signature of dynamic complexity, and its temporal correlations constrain the brain's perceptual and cognitive abilities. [4]

So statistical self-similarity (here expressed as 1/f noise) suffices for "slow oscillations" in the brain to have a causal impact on temporally embedded "faster local events". Therefore, we can cautiously say that we observe a kind of contextualization in nested rhythms which can lead to modulation in either direction – of the embedding or the embedded structure. In the acoustic case of the missing fundamental, where physically non-existent context is added inadvertently, we can find correction processes reminiscent of those at work in the corridor illusion.

The opposite phenomenon – that is to say, when our *inability* to contextualize leads to visual or auditory illusions – occurs when context is physically existent but is not taken into account in perception. An example of an underlying correction process at work which triggers an auditory illusion is the so-called Shepard scale, named after its composer Roger Shepard [5]. It consists of sequences of individual successive notes which are separated by an octave and are played over and over again. The ascending or descending sequences are superimposed but slightly staggered overlapping notes which play simultaneously and should be exactly an octave apart. Although the signal is periodic, the nested structure of the Shepard scale triggers the auditory illusion of an ever-rising or ever-falling tone. This occurs because the listener focuses on pitch relations only and thus abstracts a one-dimensional signal from a multi-layered one. To increase the illusion, each scale should fade in and out, so the listener cannot hear the beginning or the end of any given scale. A continuous version of the discrete Shepard scale, the so-called Risset scale, named after its composer Claude Risset, produces the same illusion, but with a continuous rise or fall in pitch [6]. Our inability to contextualize, and thereby create Δt_{depth}, limits us to one LOD. That is to say, we focus on pitch only, and this prevents us from generating simultaneity – the main ingredient in the construction of a perspective, as we shall see.

That we fall for such illusions is probably the result of a selection effect. Just as we benefit from our ability to identify objects regardless of distance, so do we benefit from being able to differentiate sounds only in terms of their differences in pitch. It is not necessary to have absolute hearing to register in time an approaching predator or an incoming wave: relative hearing – that is to say, perceiving changes in pitch – is perfectly adequate.

Another example of inability to perceive objects in context may suffice to show that this apparent shortcoming can turn out to be a blessing in disguise. Steven Dakin et al conducted an experiment involving the discs shown in Fig. 3.2. It will be seen that there are four isolated discs on the left side, and a disc nested into a larger one on the right side. The interiors of the discs present varying degrees of visual contrast. (The percentages on the left indicate the strength of these contrasts.)

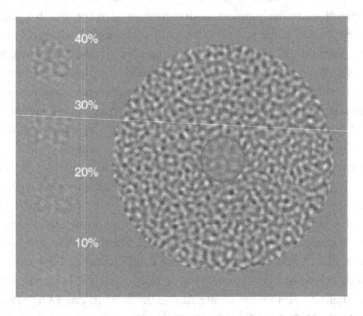

Fig. 3.2. The contrast-contrast illusion. [7] Reprinted from S. Dakin *et al.*, Weak suppression of visual context in chronic schizophrenia, *Current Biology* 15, p. R823 (2005) with permission from Elsevier.

A schizophrenic group and a control group were asked to compare the discs on the left with the embedded one on the right. They had to decide which disc appeared to contain higher contrast – the isolated or the embedded one. It turned out that individuals in the control groups tended to fall for the contrast-contrast illusion – that is to say, to them, the suppression of the contrast by its surrounding made the contrast of the embedded disc appear much lower than that of the isolated ones.

In comparison, the individuals in the schizophrenic group performed better: They did not fall for the illusion because they did not contextualize the nested disc and thus did not generate a simultaneous contrast. So their failure to use context worked – in this case – to their advantage: they succeeded in comparing the contrasts correctly, as the embedding, high-contrast disc on the right did not modify their perception of the nested one. Dakin's experiment is an example of a context which is physically there but is not taken into account in perception. We shall return to our ability to resist contextualization in Chapters 5 and 9, when we look at mirror neurons as well as nesting and de-nesting exercises and temporal misfits.

Perspective presupposes depth. This depth is generated by means of contextualizing – that is to say, by creating simultaneity through nesting. Although this may always have been true, it is only since the Italian Renaissance that we have become aware of how perspectives intrinsically work. Artistically speaking, the construction of the central perspective with its vanishing point(s) was perfected in the High Renaissance. A beautiful example is Raphael's *The School of Athens*, a fresco in the Apostolic Palace in the Vatican (see Fig. 3.3). In the centre, covering its architecture's central vanishing point, Plato and Aristotle are walking towards the spectator. Around them, other contemporary philosophers, artists and scientists are depicted.

Raphael aligns and shortens the architectural features in accordance with the central perspective and he represents the male figures in the background of the painting as slightly smaller than those in the foreground. By this means, he generates an invisible reference frame which consists of simultaneous nested layers and conveys the illusion of depth.

Fig. 3.3. Raphael's School of Athens (1510–1511). [8]

As in this spatial example, the temporal perspective also requires simultaneity of nested layers: succession alone is not sufficient to generate temporal depth. Temporal perspectives are generated by nesting multi-layered impressions [9]. This can be done by embedding retensions and protensions in the Now and nesting the present Now into future ones – that is to say, into anticipated structures, as Husserl did. The idea of nested memories is probably far easier to accept than the thought of nested expectations. But nesting expectations is exactly what we do when we interact with the rest of the world. When we intercept a tennis ball in motion, we hit it not at its perceived position, but at the point where we expect to intercept its trajectory [10]. However, all this happens within a very short moment which we perceive as one Now. So we may say that the Now which spans the perception of the ball's position is nested into the slightly longer Now, in which we see the ball, take a big swing and intercept its flight with our racket.

Such anticipatory behaviour is the result of adaptation through training. If the ball comes through faster or slower, a novice takes longer than a professional to adapt his interceptive movements. So just as tuning spatial perspective is an acquired skill – one which Kenge the young Pygmy had not fully mastered – so temporal perspectives also require adaptation and continuous retuning through training. Building on extensive work on anticipatory systems done by Rosen [11] and Dubois [12], we shall take a closer look at anticipation and compensation for time delays in the next Chapter. For now, though, it shall suffice to think of anticipation as a means of nesting the present Now into not-yet-existing future ones.

Most of us can hit a ball, hear a tune and enjoy both temporal and spatial depth, unless we suffer from sensory deprivation, a neurodegenerative disease or are otherwise impaired. In general, we may say there two types of observers. Those with a fractal, or nested, perspective enjoy both temporal and spatial depth. Those who lack the ability to nest are limited to the perception of successive events on one LOD. Henceforth, I shall refer to the former group as fractal observers and the latter as non-fractal ones. Non-fractal observers would not be able to hear a tune because they are not capable of producing nestings – a prerequisite for creating simultaneity – that is to say, Δt_{depth}. Neither would they be able to correctly estimate the projected size of figures in a painting arranged in a central perspective. Luckily, most of us are fractal observers and can enjoy listening to a tune or seeing a forest rather than a series of trees.

So far, we have dealt with compatible nested perspectives only: Raphael's central perspective was compatible in the sense that its spatially nested layers could be perceived simultaneously and interpreted completely and without contradictions. An example of an incompatible nested perspective is Holbein's painting *The Ambassadors* (Fig. 3.4). It contains an anamorphosis – that is to say, a distorted perspective – which an observer can interpret fully only if he looks at it from two observer perspectives: the central perspective and another which deviates significantly from the normal observer standpoint.

Fig. 3.4. *The Ambassadors* by Holbein (1533). [13]

This second perspective, which he may attain by moving to the left and ducking, reveals an anamorphosis in the shape of a *memento mori* – a Latin term meaning "remember that you are mortal" or "remember that you will die". From the central perspective, the highly distorted skull which Holbein inserted into the painting cannot be made out as such. (In Holbein's time, such play on perspective which questioned the reality and truth of our visual perceptions enjoyed great popularity.)

Once the observer has adapted to the fact that he is dealing with two incompatible nested perspectives, he can make out the skull even from a central perspective. The process of becoming aware of a distorted subset within the painting has modified the structure of his way of perceiving the world. It is very likely that the next time this observer is confronted with a similarly distorted subset, he will detect the incompatible perspectives [14].

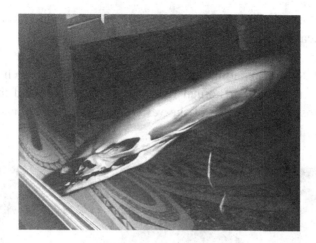

Fig. 3.5. The distorted skull. [15]

Temporal incompatible perspectives are harder to identify. Earlier in this chapter, I mentioned the Shepard scale as an example of a nested sequence of tones. Although listeners first fall for the auditory illusion of an ever-ascending tone, they may be trained to hear the periodical sequences instead. To play the notes at the start of the sequence louder would facilitate the task of focussing not on pitch but on the first note of the ever-repeated overlapping series. This way, we can train ourselves to choose which of two incompatible perspectives we want to hear: the periodic sequence or the auditory illusion.

Incompatible perspectives cannot always be identified as such and conflicting expectations may seriously incapacitate an observer dealing with such a clash. This may occur if he is stuck with a temporal perspective because he is not in a position to analyze his situation by determining the number of nestings his perspective holds – that is to say, if he cannot determine Δt_{depth}. Prolonged exposure to incompatible perspectives may create a temporal pathological state, in which a petrified observer is unable to interact with the world. An example of such a rigid perspective is a double bind relation. A double bind is a behavioural pattern in which one speaker sends out conflicting verbal and/or nonverbal messages to another, who cannot respond because the conflicting messages make him unable to [16]. In Chapter 9, we shall take a closer look at contradictory perspectives which can lead to pathological states.

We are usually not aware of the fact that our perspective is made up of nested Nows. Sometimes, unexpected perspectives or even linguistic curiosities interrupt our normal flow and make us stop and think. One such curiosity is the name of the *yesterday-today-tomorrow* shrub (see Fig. 3.6).

Fig. 3.6. The yesterday-today-tomorrow shrub. [17]

Within a few days, the colour of its blossoms changes from deep violet-blue to a light blue and finally to white. As this slow fading of the colour is staggered, for a longish period of time the plant displays blossoms of all three colours. Now, in order to come up with a name like *yesterday-today-tomorrow* shrub, whoever coined its name described the blossoms on at least two LODs – that is to say, the plant as a whole and its individual blossoms. By taking both a global and a local view simultaneously, the name-giver referred to the past, the present and the future in one Now. To achieve this, it takes a fractal observer with a perspective made up of nested Nows, which provides for both simultaneity and succession. A non-fractal observer would have looked at the individual successive colour change of the blossoms on one LOD only. Although the scaffolding of our temporal fractal perspective is invisible, it is possible to deduce its existence from examples such as this naming of a shrub or the anamorphosis we saw in the Holbein picture.

There are both internal and external constraints on our spatio-temporal perspectives. The latter manifest themselves as simultaneity, succession, duration and the Now. None of these notions are absolute concepts. As we have seen in Chapter 1, duration is highly context-dependent. Simultaneity and succession are also relative notions which do not make sense unless an observer perspective is specified. The truth value of statements about temporal succession – for instance like "A happens before B" or "A is simultaneous with B" – are also context-dependent.

An example of the context-dependence of simultaneity comes from Einstein's Special Theory of Relativity, which defines these contexts as inertial systems. Inertial systems can be regarded as abstractions of concrete physical observers whose measurement instruments are at rest relative to each other and who are not subjected to any forces. Two different inertial systems may be in relative constant motion. These restrictions are lifted in General Relativity which deals with accelerated systems and systems that are exposed to gravitational fields. However, in order to describe the relativistic concept of simultaneity, the Special Theory is sufficient.

Einstein defines two spatially separated events E_A and E_B taking place at locations A and B as simultaneous if they are both triggered by a flash of light emitted at the mid-point M between A and B. This definition of simultaneity is quite intuitive if the respective observer is at rest relative to the locations A and B. A moving observer, however, sees things differently: For him, locations A and B are moving and their distances to the location of the light emission are no longer identical at the times of the events E_A and E_B. Together with the assumed invariance of the speed of light, this leads to an asynchrony of E_A and E_B for the moving observer. Depending on the direction of the relative motion, the moving observer will perceive event E_A before E_B or vice-versa.

For our purposes, it will suffice to acknowledge that simultaneity and the order of events are inextricably linked to the observers' relative motions which shape their perspectives. Einstein's introduction of the concept of observer frames meant that there is not absolute simultaneity of two events if they are spatially separated. All we can talk about is simultaneity in a relativistic and therefore context-dependent interpretation. This is an important observation, as it shifts – on the physical level of description – the multi-levelled simultaneous layers which make up the observer's perspective into the eye of the beholder.

But there are other constraints on simultaneity – both physical and psychological – which we encounter in our everyday lives on planet Earth, when we synchronize various sets of multi-layered events in our Now. One constraint is an external one, which is linked to the fact that certain signals which reach us propagate at different speeds. If a thunderstorm is sufficiently far away from us, we first see the lightning and then hear the thunder, because light propagates faster than sound. Another constraint – an internal one – is determined by the temporal threshold beyond which we perceive two successive events as one because the interval is too short for our perceptual apparatus to differentiate. We perceive auditory signals as non-simultaneous – that is to say, as successive, if they are separated by an interval of approx. 6 milliseconds. Below this threshold, we perceive them as being simultaneous. Tactile impressions merge into one event at a threshold of roughly 10 milliseconds. Visual impressions, on the other hand, are perceived as simultaneous if they are separated by an interval of 20 to 30 milliseconds. So if we experience sound, vision and touch at the same time, we may say that we are dealing with three nested simultaneous perceptions: the auditory perception is nested into the tactile, which, in turn, is nested into the more extended visual Now (see Fig. 3.7).

This incongruence is smoothed out by another corrective distortion we perpetually perform without noticing. Although we react to acoustic stimuli faster than to visual ones, our brains integrate the various sensory inputs into one gestalt. The German neuroscientist and psychologist Ernst Pöppel defines a simultaneity horizon, which spans at a radius of approximately 10 meters from the observer. At this distance, we perceive audio and visual stimuli as being simultaneous. This is because the difference in the speed of light and sound is compensated by the difference in our reaction time to visual and acoustic stimuli. Within the simultaneity horizon, we perceive acoustic stimuli before visual ones. Beyond it, visual stimuli reach us before acoustic ones [18].

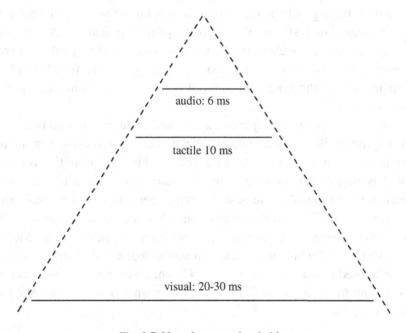

Fig. 3.7. Nested sensory thresholds.

American neuroscientist David Eagleman, who specialises in the field of time perception, particularly in the distortion of duration and temporal order, also looks at how the brain synchronizes multi-layered, multi-modal incoming signals [19]. Like Pöppel, he shows that the brain takes account of speed disparities among and within its sensory channels when it tries to construct a coherent world by waiting for the slowest information to arrive. Then, within that window of delay, it makes sense of the gathered data in retrospect or, as Eagleman puts it, postdictively. An example of how the brain co-ordinates incoming multi-modal signals is the firing of a gun to start a race. A spectator close to the starting line sees the firing and hears the bang simultaneously although vision and hearing process information at different speeds. The corrective distortions our brain performs to create a unified sensory perception of the world are remarkable. Just as we create size constancy – an example of feature-binding, i.e. our ability to recognize objects under varying conditions – by means of a corrective distortion, so do we perform temporal-binding when we assign a certain order and duration to perceptions. For instance, if you touch your nose and toe at the same time, you feel the touches simultaneously, although the signal from your nose reaches your brain faster than that from your toe. We shall take a detailed look at the synchronization of perceptions and temporal binding in Chapter 6.

As we saw in the example of the Shepard scale, non-nested hearing is an acquired skill: resisting contextualization in order to avoid hearing the auditory illusion requires practice (and a little help in the form of emphasizing each beginning of the periodic signal). As adults, we have learned to contextualise, so nested hearing seems to be our normal mode of perception. However, it turns out that the skill of nesting our perceptions, even of one mode, is also not an innate skill, but an acquired one. Christian Tschinkel, an Austrian musicologist and composer recalls how he used to listen to his parents' 45 rpm single record which seemed to contain the same piece (Richard Clayderman's *Ballade pour Adeline*) on both sides:

> Approximately at the age of four, I often compared the A and B sides without being able to make out any kind of difference concerning the music. As other single records contained different songs, I rated this as a pure waste of space. Only a few years later did I realize that one side contained a version for solo piano and the other a version for orchestral accompaniment. This example shows that, as children, many of our perceptional performances are still very limited and we thus establish a completely different world [20].

Nested hearing – that is to say, the perception of multi-layered signals of various extensions – is also an acquired skill, like all contextualizations and corrective distortions our brain has learned to perform. The resulting increase in available contextualisations has been described by developmental psychologist H.H. Clark in terms of the learning child's continuous acquisition of rules of application [21]. Through interacting with the world, a child acquires a large number of new LODs. Often, however, he will find himself confronted with a situation he cannot comprehend by means of the metaphors already at his disposal. In such cases, he would either not respond or try to use metaphors he is already familiar with. Clark cites the example of a small boy who, when looking at a picture book, was asked "When did the boy jump the fence?" Pointing to the fence in the picture, he answered "There!". That is, the boy had chosen to apply a spatial metaphor familiar to him, rather than a temporal one which he has not yet mastered [22].

Building on such observations, Clark formulates the *complexity hypothesis* which is based on the correlation between human levels of perception and the appropriate language levels [23]. It states that the order in which spatial concepts are acquired is imposed by rules of application which include direction, point of reference and dimension. If, of two terms A and B, B requires all the rules of application A requires plus an additional one, then A is acquired before B.

An example is the order in which the prepositions *in*, *into*, and *out of* are acquired: *in* presupposes a three-dimensional space; *into* presupposes a three-dimensional space and a positive direction – that is to say, in the direction of the stronger perceptual field and *out of* presupposes a three-dimensional space, a positive direction and a negation of this direction. So we acquire first *in*, then *into*, and finally *out of*, since the second and third prepositions each require all the rules of application of the preceding one(s) plus one in addition.

The complexity hypothesis predicts for spatial and temporal terms the following order of acquisition: In antonymous pairs, the positive term will be learned before the negative one (e.g. *into* before *out of*). *At*, *on* and *in* are acquired before *to*, *onto* and *into*, because the latter require an additional direction. Location prepositions such as *at*, *on* and *in* are mastered before correlative location prepositions such as *above* and *in front of*, as the latter require a point of reference in addition. *Tall* and *short* will be learned before *thick* and *thin*, because the latter requires an additional dimension. Unmarked neutral terms which lack connotation will be acquired before marked neutral terms loaded with (usually negative) connotation. Positive terms are acquired before their negative opposites, with the positive term determining the dimension: *long* (+), *short* (-) $\Rightarrow$ dimension: *length*. Every additional rule of application adds another layer – that is to say, another LOD. New LODs provide a new contexts by adding, for instance, a direction or a connotation, which make the term more complex. As this form of complexity retains the earlier acquired, simpler terms, we can say that the simpler terms are nested in the more complex ones – a fractal arrangement.

Against the background of a fractal temporal perspective [24], learning is a result of generating new rules of application in the form of LODs. By contextualizing these LODs with every new experience or recollection, we continue to nest them into a cascade of LODs, which each time leads to a further increase in Δt_{depth} in the Now. This has an effect on our perception of time: When we have generated a large number of LODs, we are in a position to arrange events in nestings, i.e. in "during-relations" rather than having to successively arrange them, like beads on a string, on one LOD. This could explain why a summer of one's childhood is so incomparably much longer than a summer of one's adulthood.

The act of recollecting may be regarded as learning too, as nesting means arranging past events in a new context. Such nestings by recollection often occur in clusters, when a number of people discuss past events they remember. Examples of typical recollection clusters comprise class reunions, weddings, anniversaries, housewarming parties and family slide-shows on Christmas Eve. Think of a class reunion, where recollections continue to be nested and rearranged, as old stories are discussed, corrected and retold by the former classmates. As a result of this continuous increase of simultaneity, Δt_{depth} increases perpetually and Δt_{length}, by contrast, contracts steadily. For the former classmates, time flies. For the pitiable accompanying husbands and wives, Δt_{length} increases steadily, as they cannot join in the recollecting and have to arrange every more or less insightful remark of the former classmates on a constant number of LODs – in other words, they are bored stiff.

I have described this familiar phenomenon of subjectively varying perception of duration in Chapter 2 and suggested that the distribution of Δt_{depth} and Δt_{length} accounts for temporal dilations or contractions. Both these measures plus the resulting fractal dimension are measures for comparing the temporal complexity of processes. When we use the notion of complexity, we usually have some vague idea of what we mean by it. In everyday language, we may equate complexity with complicatedness. However, it is worth taking a closer look at this concept, since it reveals possibly unreflected assumptions we maintain about absolute and relative order.

In Clark's hypothesis, the notion of complexity refers both to the degree of difficulty of applying rules as well as the intrinsically complicated structure of those rules themselves. There are at least as many senses of the word complexity as there are human faculties, some relating to the level of internal organization of a structure, others to the difficulty of describing or constructing a structure. Some measures of complexity are, at first sight, defined by their intrinsic organization. Algorithmic complexity is such a case. It uses the length of the most compact description of an object as a measure of complexity. If we compare the two strings "XXYXXYXXYXXYXXYXXY" and "LQHSIHGUBVAPKZY", the first one seems to be less complex because we can formulate a simple generation rule. The second string cannot be compressed, so its shortest description is the string itself, which means it is random. John Casti describes the algorithmic complexity of an object as "directly proportional to the length of the shortest possible description of that object." [25] He further observes that

A string of letters is random if there is no rule for generating it whose statement is appreciably shorter – that its, requires fewer letters to write it down – than the string itself. So an object or pattern is random if its shortest possible description is the object itself. Another way of expressing this is to say that something is random if it is *incompressible*. This idea forms the basis for the theory of algorithmic complexity (...).[26]

So, by this measure, the highest degree of complexity may be equated with randomness. Other measures of complexity focus on completely different properties: the fractal dimension, entropy, information, correlation, grammatical complexity, hierarchical complexity – there is an abundance of measures which all use the same term but refer to very different properties. And although, as Casti points out, an amoeba is less complex than an elephant whatever notion of complexity we subscribe to, most of these measures make sense only if the complexity of the observer is also taken into account.

This puts complexity – just like fractality – into the eye of the beholder. Your internal degree of complexity and the resulting structure of your observer perspective provide the context in which you recognize a structure or a process as simple or complex. Apart from the code, the medium of transmission and the message, context plays an essential role in the generation of complexity:

> As a result, the complexity of a political structure, a national economy or an immune system cannot be regarded as simply a property of that system taken in isolation. Rather, whatever complexity such systems have is a joint property of the system and its interaction with another system, most often an observer and/or controller. [27]

A further definition of a measure of complexity which explicitly takes account of the observer-context relation will be introduced in the next chapter. It measures only the difference between the number of internal and external nestings and takes into account whether or not the observer is aware of those nestings.

In this chapter, I have suggested that an observer constructs a perspective by generating depth. He creates an invisible reference frame by representing objects at different spatial or temporal distances from the observer as having different extensions. A temporal perspective is generated by the nesting of multi-layered signals and reiterated retensions and protensions – that is to say, the nested Nows. Perspective is a relative ordering structure which presupposes simultaneity of different nestings. If we observed the contents of different nestings successively or – as Powers puts it – if time were "just one damn thing after another" [28], no perspective would arise as no Δt_{depth} would be generated. We are nested observers in a nested world.

We have looked closely at the structure of our temporal perspective – that is to say, our Now, which is our only window to the outside world – and suggested that corrective distortions are adaptations to our present environment and anticipated future. Temporal corrective distortions are ubiquitous but reveal themselves only when our expectations are not met, as in auditory illusions or modified judgements of duration or temporal order. They may also be present, in the form of delays or rhythmic mismatches, in disorders which result from problems of timing both within the human brain and body as well as between the individual and his context. These issues, and the question of the role fractality might play, will be discussed in Chapter 9.

We have talked about constraints on our perception and limitations to our perspectives. Many of these constraints arise from the fact that, as we shall see in Chapter 4, internal and external perspectives display different levels of complexity. The difference in the amount of visible nestings is the boundary which connects us to and separates us from the rest of the world and which defines self and non-self. Let us now take a closer look at the structure of this boundary, which shapes our temporal perspective and perceptual event horizon.

References

1. C. M. Turnbull, Some observations regarding the experiences and behaviour of the BaMbuti Pygmies. *American Journal of Psychology* Vol. 74, 304-308 (1961).
2. Illustration courtesy of Syed Azaharul (2009).
3. E. Husserl, (1928); *Vorlesungen zur Phänomenologie des inneren Zeitbewußtseins.* (Lectures on the phenomenology of the inner consciousness of time.) First published in 1928. Niemeyer (1980).
4. G. Buzsáki, *Rhythms of the Brain.* Oxford University Press, Oxford, U.K., p. 134. (2006).
5. R. N. Shepard, "Circularity in Judgements of Relative Pitch" *Journal of the Acoustic Society of America* **36** (12): 2346-53, Dec. 1964.

6. C. Tschinkel, W. Musil, The sound files of the Risset scales (rising, descending and a dichotic version of rising and descending scales) were kindly provided by Christian Tschinkel and Wolfgang Musil (2006). http://tschinkel.hof-productions.com.

7. Reprinted from S. Dakin, et al, Weak suppression of visual context in chronic schizophrenia. *Current Biology* 15, p. R823 (2005) with permission from Elsevier.

8. Photograph courtesy of Roland Loser 2010.

9. S. Vrobel, "Simultaneity and Contextualization: The Now's Fractal Event Horizon". Talk held at *13th Herbstakademie: Cognition and Embodiment*. Ascona, Switzerland (2006).

10. N. Benguigui, R. Baurès and C. Le Runigo, Visuomotor delay in interceptive actions, in: *Behavioural and Brain Sciences* No. 31, p. 200 (2008).

11. R. Rosen, *Anticipatory Systems: Philosophical, Mathematical and Methodological Foundations*. Oxford, Pergamon Press (1985).

12. D. M. Dubois, Introduction to Computing Anticipatory Systems, *International Journal of Computing Anticipatory Systems*, CHAOS, Liège, Belgium, Vol. 2, pp. 3-14 (1998).

13. Photograph courtesy of Amrit MacIntyre.

14. S. Vrobel, "Temporal Observer Perspectives" in: *SCTPLS Newsletter*, Vol. 14, No. 1 Society for Chaos Theory in Psychology & Life Sciences, October 2006.

15. Photograph courtesy of Namaby.

16. M. Koopmans, From Double Bind to N-Bind: "Toward a New Theory of Schizophrenia and Family Interaction", in: *Nonlinear Dynamics, Psychology and Life Sciences*, Vol. 5, No. 4, pp. 289-323 (2001).

17. Photograph courtesy of Judith Ridge.

18. E. Pöppel, *Grenzen des Bewußtseins – Wie kommen wir zur Zeit, und wie entsteht die Wirklichkeit?* Insel Taschenbuch, Frankfurt, pp. 38-42 (2000).

19. D. Eagleman, The Eagleman Lab: http://:neuro.bcm.edu

20. C. Tschinkel, "Perception of Simultaneous Auditive Contents", in: *Simultaneity – Temporal Structures and Observer Perspectives*; (eds. S. Vrobel, O. E. Rössler and T. Marks-Tarlow): World Scientific, Singapore, pp. 3-14. p. 373 (2008).

21. H. H. Clark, "Time, Space, Semantics and the Child", in: T. E. Moore (Ed.): *Cognitive Development and the Acquisition of Language*. Academic Press, London, 1973.

22. Ibid.

23. Ibid., p. 54.

24. S. Vrobel, "How to Make Nature Blush: On the Construction of a Fractal Temporal Interface" in D. S. Broomhead, E. A. Luchinskaya, P. V. E. McClintock and T. Mullin (Eds.), *Stochastics and Chaotic Dynamics in the Lakes: STOCHAOS*. New York: AIP (American Institute of Physics), pp. 557-56 (2000).

25. J. L. Casti, *Complexification – Explaining a Paradoxical World Through the Science of Surprise*. HarperPerennial, New York, p. 9 (1995).

26. Ibid.

27. Ibid.

28. R. Powers, *The Time of Our Singing*, Vintage, p. 95 (2004). In common with other commentators, Powers has here adapted Arnold Toynbee's celebrated definition of history.

Chapter 4

The View from Within: Extended Boundaries

Our spatial and temporal perspectives are determined by the way we distribute simultaneous and successive nestings. In this chapter, we will first take a look at the structure of the boundaries which connect and separate such nestings. Then, we shall investigate to what extent shifted boundaries can shape our perceptions and, therefore, our temporal perspectives. Next, we shall distinguish between visible and invisible interfaces and devise a measure of complexity which compares internal and external perspectives. Internalized models will be described as anticipative systems. Finally, we discuss how different cuts produce different temporal perspectives and discuss models and methods for identifying boundaries.

If you have ever taken a holiday in the Tyrolean Alps, you have probably noticed that almost everyone, whether venturing to climb a nearby peak or just aiming for the next beer garden, is carrying a walking stick. These sticks are typically wooden, with a curved handle and capped by a metal tip at the end. Frequent use of such a stick can turn it into an extension of the arm which carries it. Such a walker can often be seen directing his stick smoothly along the ground and striding fluently, almost effortlessly. The walker himself will notice a change in his proprioception: to him, it may seem as if he has grown an extra limb at the end of which he now touches the world. We may say that there are two key interfaces at work here: the first is located at the spot where the hand grasps the stick, the second one where the now extended observer – consisting of walker + stick – touches the ground [1].

By incorporating a part of the world which was previously not assigned to our bodies but to our environment (when the stick was still in the shop window), we form a new systemic whole. A systemic whole is defined by the relationship of its elements, which we experience as a unity. In particular, "one of the significant characteristics of a system of this type is that there are properties of the whole that cannot be found in the elements." [2] This definition implies that the creation of a new systemic whole is an emergent phenomenon – that is to say, the whole is more than the sum of its parts. And the elements – the two interfaces in our example – are experienced as one whole, namely, the boundary into which those interfaces have merged. Successful incorporation of a tool like a walking stick is perceived from the outside as fluency in movement and from the inside as a merging of interfaces. One effect of this merging is transparency: the individual components of the extended system walker + stick have become invisible to the walker when, through frequent use, they are fused into one systemic whole [3].

Fig. 4.1. A new systemic whole: walker + stick. [4]

Incorporating a tool is a method used, for instance, in robotics, when adaptive agents – that is to say, robots which are capable of interacting with their environment and learning – are trained to develop smooth and effortless movements. It is also exploited in therapy, when individuals suffer as a result of assigning parts of their bodies or minds to their environment and no longer recognize them as their own. An example of such misattribution is the condition known as unilateral neglect, when someone sees only one hemisphere of his normal visual field or is convinced that, for instance, one of his extremities does not belong to his body. Such an individual selectively ignores a part of his egocentrically encoded space. For instance, he may insist that his left leg is not his – a conviction that does not change even if relatives or doctors point out to him that it is firmly connected to his pelvis [5]. Another may be completely unaware of the fact that he has shaven only the left side of his face or has eaten only the food on the right half of his plate. When such individuals, who are blissfully unaware of the existence of one hemisphere, are trained to use, for example, a stick as a tool for reaching, their visual field can magically extend into the space which was hitherto neglected but which has now become reachable by means of the tool. The unexpected access to the neglected hemisphere is a result of the fact that our brain distinguishes between *far space* (ie beyond reaching distance) and *near space* (ie within reach). When we get used to employing a stick for reaching, we remap far space to near space – that is to say, a part of the environment is incorporated when the brain interprets the stick as an extra limb. We may, at this point, truly talk of an incorporation – as opposed to a mere accommodation – because the process of assimilating the tool is accompanied by plastic neural changes [6].

And since we are embodied individuals who are not locked-in but flexibly interact through our bodies and minds with our environment, we constantly re-negotiate the boundary between inside and outside, between self and non-self. As we have seen, this recalibration process can go both ways – we can become extended or reduced observers, depending on whether we assign parts of our environment to ourselves, as in the stick example, or vice-versa, as happens when a hemisphere of visual space is neglected.

Another example of how negotiable our boundaries are is the famous rubber hand illusion. Botvinick and Cohen [7] conducted an experiment in which the participants placed one hand on a table in front of them and the other underneath that table. Then a fake hand was put in the position where the hidden hand would have been resting. The participants were asked to look at the fake hand while both that hand and the out-of-sight real hand were stroked with a brush. But, after a while, only the fake hand was being stroked, while the real hand was no longer being touched at all. Still, participants reported feeling the same brushing sensation on their real, hidden hand while they observed the fake one being stroked. The visual feedback alone was sufficient to maintain the illusion of touch. The rubber hand illusion is an example of a fast, yet ephemeral assignment of a part of our environment to our bodies: It only takes a few minutes to evoke the illusion that the rubber hand belongs to the body, but it is maintained through visual feedback only for a short while.

Such distortions in proprioception – be it in the shape of reductions or extensions of the observer's sphere of influence – may be interpreted as boundary shifts. The failure to perceive a hemisphere of visual space shifts the boundary between the observer and the rest of the world (which I shall henceforth refer to as world-observer boundary) towards the inside: we may say that the observer's extension is reduced. By contrast, the rubber hand illusion shifts that boundary outwards when, for a short time, a new systemic whole (human being + fake hand) comes into being.

Besides spatial observer extensions such as those portrayed above, there are also temporal ones. They manifest themselves in nested rhythms both within the body and in the environment. In Chapter 2, I briefly introduced the idea of nested rhythms in the shape of neural oscillations which are nested into slower metabolic ones, both of which rhythms are again nested in tidal, seasonal and astronomical ones. Buzsáki points out that both action and cognition are based on the brain's ability to generate and sense temporal information:

> This temporal information is embedded in oscillations that exist at many different time scales. Our creativity, mental experiences and motor performance are modulated periodically both at short and long time scales. [8]

What emerges is a fractal temporal structure, generated by nested oscillations, which are sometimes triggered and maintained by external stimuli. At other times, they are self-organized, which means the spatio-temporal structure evolves spontaneously in the absence of external influences. (Note that I follow Buzsáki in using the terms rhythm, oscillation and periodicity interchangeably. They describe the same phenomenon, but the individual terms have been adopted by different academic disciplines [9].) If internal and external rhythms are coupled successfully, the parts of the temporal environment governed by those rhythms can be said to have become incorporated in the observer.

The internal rhythms which drive the dynamics of our bodies require frequent recalibration. This becomes necessary whenever we encounter new contexts or have to adapt to a changing environment. As, among others, van Nieuwenhuijze has pointed out, successful nesting of internal rhythms into external ones as well as well-calibrated internal oscillations are prerequisites for good physical and mental health [10]. Internal rhythms interact both on the same LOD (when we look, for instance, at communication among cells) and hierarchically (that is to say, between nested levels). An example of such hierarchical or "vertical" nesting is the communication between cells, organs, the entire body and the social environment. As both our bodies and our environment are in constant flux, the art of tuning those nested rhythms into a harmonious symphony is a constant balancing act. Through nesting internal and external oscillations, we can create a temporally extended observer.

Some internal rhythms have become incorporated to such an extent that no external timer is required to maintain this periodicity, although in terms of evolution, the internal rhythms have originally evolved by adapting to external ones. One example is our circadian rhythm – that is to say, our sleep-wake cycle, which no longer requires the external pacemakers of daylight and darkness. When people are exposed to exogenous influences such as sunlight and social contacts, their circadian rhythm adopts a 24-hour cycle. However, if individuals are deprived of such external pacemakers, their circadian rhythm is shortened by an hour but levels at around 25 hours [11]. The ability to maintain internal rhythms which are not perturbed by environmental changes is a great advantage and, therefore, a selection effect.

Most contextual change, however, requires us to adapt. Clark reports an experiment which provides an impressive example of how living beings adjust to environmental changes by successfully re-calibrating their temporal interfaces. A macaque monkey found himself confronted with delay times which were added to or taken away from a course of actions he had learned to perform [12]. The monkey had learned to move a cursor on a computer screen by means of a joystick, while the computer recorded the neural activity which accompanied those actions. Then the joystick was disconnected from the computer. The monkey continued his task, still manoeuvering the now detached joystick while he was actually moving the cursor by means of his neural activity. In the next step of the experiment, a robot arm was inserted into the control loop. This mechanical device, which added a slight delay to the cursor-steering process (because of the mechanical friction it produced), could not be seen by the monkey. The addition of the delay, which the monkey noticed in the visual feedback – that is to say, in the delayed motion of the cursor on the screen – completely confused it at first. But after this, it adapted to the new situation and made up for the delay. In other words, the monkey's temporal interface was extended when it compensated the delay by means of anticipatory regulation – that is to say, when it "skipped" the consecutive interfaces in the control loop and fused them into one, as does the walker when he incorporates his stick. The monkey never suspected that it moved the robot arm and thus the cursor by thought control alone – this was only clear to the external observers, namely, the individuals conducting the experiment. To the monkey, the merged interfaces (neural records / robot arm / cursor) were not visible individually – they formed *one* boundary. To an external observer, by contrast, the couplings of the individual interfaces were visible.

The view from the outside allows us to see both the detailed successive interfaces and the one boundary these interfaces have merged into. The examples above show that whenever a new systemic whole comes into being, be it an extended or reduced version of the original whole, the world-observer boundary shifts either towards the inside or the outside. New systemic wholes are formed when conditioning effects make these boundaries invisible to the observer, as is shown in the example of the walking stick which is assimilated into an extra limb. An external observer would be able to make out two interfaces: one where the hand touches the handle of the stick and a second one, where the walking stick touches the ground. The view from inside the systemic whole walker + stick reveals only one interface which connects the hand directly with the ground. The walker is no longer aware of the now merged interfacial chain.

This difference between the inside and outside view is a useful measure of complexity Henceforth, I shall, after Otto Rössler, refer to the inside and outside vantage points as endo- and exo-perspectives [13]. One way of quantifying and comparing boundaries in terms of their spatial and temporal extensions is to simply count the number of merged interfaces which are no longer visible from within the systemic whole but can be made out from the exo-perspective – the vantage point of an external observer, and compare it to the number of interfaces at work from an endo-observer's perspective. The difference between the outside and inside perspectives is the number of visible assimilated or rejected contexts, which can be measured in Δt_{depth} – that is to say, the number of nested interfaces. This relation between Δt_{depth} (endo) and Δt_{depth} (exo) I denote as the *boundary complexity* of an embedded endo-system. So, in the walking stick example, the exo-perspective reveals two interfaces, which have merged into one for the endo-observer:

$$\frac{\Delta t_{depth} (\text{endo}) = 1}{\Delta t_{depth} (\text{exo}) = 2} = 0.5$$

which gives a value of 0.5 for the boundary complexity [14]. If we speak in terms of temporal nestings, the observer becomes a new extended systemic whole if he successfully couples or assimilates internal and external rhythms and thus incorporates parts of his temporal environment. Alternatively, he could also become a new reduced systemic whole, when he assigns parts of himself to his environment or suddenly fails to contextualize embedded patterns and react as the schizophrenics in Dakin's experiment (see Chapter 3). If a delay is newly integrated into an observer's sphere of influence, he will adapt through regulatory anticipation, and turn into a temporally extended observer. If an already incorporated delay is shortened or removed from his sphere of influence, an observer will adapt through retardation and turn into a temporally reduced observer. Our temporal recalibration can be remarkably persistent – a fact dramatically shown by our distorted perception of events when delays we have adapted to are suddenly removed or shortened. Stetson et al describe an experiment in which a consistent delay was inserted between the participants' pressing of a key and a flash which invariably followed it [15]. After they had adapted to the delay – that is to say, when they had incorporated it, participants were faced with shorter delays. Surprisingly, this made them perceive the flashes as occuring *before* the pressing of the key – a temporal distortion in which action and perception were reversed because a delay was shorter than anticipated. Thus, we may deduce that temporal boundary shifts can distort our perception and the temporal order of our nested Nows.

But, returning to our notion of a personal temporal perspective, can we actually say that a temporally extended observer's Now is expanded and that a temporally reduced observer's Now has shrunk? We can indeed, provided the nested interfaces have merged and are no longer visible to the endo-observer. A fully incorporated delay can no longer be separated from the whole action it is embedded in. Only an exo-observer is able to divide the new systemic whole into its temporal components. The idea is reminiscent of French philosopher Henri Bergson's notion of internal duration (*dureé*) as an indivisible whole:

> ... picture the image of an infinitely small elastic band, contracted, if it were possible, into a mathematical point. We slowly start stretching it, so that the point turns into a line which grows continuously. Let us focus our attention not on the line *qua* line, but onto the action of pulling it. Notice that this action is indivisible, given that it would, were an interruption to be

inserted, become two actions instead of one and that each of these actions is then the indivisible one in question. We can then say that it is not the moving action itself which is ever divisible, but the static line, which the action leaves under it as a trail in space. [16]

The action of pulling the elastic band is an indivisible whole for the endo-observer who performs this action. The internal duration of this whole is a meaningful entity which spans the endo-observer's Now. Incidentally, Bergson also realized that duration requires a nested structure of our Now when he talks about the present as containing a perpetually expanding image of the past:

> The internal duration is the continuous life of a recollection which extends the past into the present, so that the present may clearly contain the perpetually expanding image of the past ... . Without this continuing existence of the past in the present, there would be no duration, only the existence of the moment. [17]

Systemic wholes display extended, yet – from an endo-perspective – indivisible temporal perspectives. The internal fractal structure in the shape of nested detail becomes invisible to the endo-observer when those nested interfaces have merged into one boundary.

We may say that for an endo-observer, his world-observer boundary expands and is shifted outwards when it incorporates a model of the observer himself and his environment. An exo-observer also uses models, but he does not incorporate them – they are part of his environment and thus beyond his world-observer boundary. In the terminology of anticipatory systems theory, an endo-observer who has constructed or incorporated a model of himself and his environment displays what Dubois denotes as *strong anticipation* or *endo-anticipation* [18]. (Dubois defines an anticipatory system as "a system that computes its current states in taking into account not only its past and present states but also its potential future states" [19].) *Weak anticipation* or *exo-anticipation*, by contrast, merely requires an observer to use a model of the external system for which he intends to make a prediction. An example of exo-anticipation is weather forecasting by means of a computer simulation. An example of endo-anticipation is human planning – here, the future aim, which may be represented, for instance, by a diary entry marking the date of my birthday party, determines the action which temporally precedes and leads to the envisaged event (contacting friends, writing invitations, shopping for food and drink, etc). In a way, we may say that, in this case, the future event causes my actions in the present: Because I have announced a party which will take place in a month's time, I am now sitting at my desk writing invitations. This type of purpose-driven causation is also known as final causation or *causa finalis*. This notion goes back to Aristotle, who formulated four types of causation: formal, material, efficient and final cause. The formal cause (*causa formalis*), the material cause (*causa materialis*), and the efficient cause (*causa efficiens*) explain what, out of what and why something is happening in the present as a result of past events. The final cause (*causa finalis*), on the other hand, is goal-directed: it explains to what purpose something is happening in the present, so it is the future which determines the course of events. Today, final causation plays an essential part in Dubois' anticipatory systems, which pay tribute to both the *causa efficiens* and the *causa finalis* by posing both questions: as a result of what and to what purpose does the present state arise?

Both types of causal relations are also at work in nested systems in which memory and anticipation determine present decisions (as was the case in Husserl's example of how we manage to perceive a tune. See Chapter 2). My Theory of Fractal Time, the first part of which I briefly presented in Chapter 2, also attributes causal relations in Δt_{length} – that is to say, between successive events, to both the *causa efficiens* and the *causa finalis*. This is so because Δt_{depth} logically precedes Δt_{length}: we cannot conceive of successive events unless they are arranged against the background of an embedding, or overlapping simultaneous event. An example of an embedding event can be found in Buzsáki's observation that global slow oscillations modulate faster local events. Another example would be the case of the missing fundamental, where our perception of nested, high frequencies triggers the emergence of embedding, lower ones (both presented in Chapter 2). Overlapping events occur when we remember the past and anticipate the future in our Now, so retensions and protensions merge into one nested temporal structure. Strong or endo-anticipation implies that longer temporal intervals can influence embedded shorter ones and that shorter intervals can influence longer, embedding ones. Along with these directions of causation from the inside to the outside, and vice-versa, come boundary shifts which create new systemic wholes by extending or reducing the original system (for instance, a walker using a stick or a patient for whom a spatial hemisphere is lacking). In this context, it is interesting that for Dubois, anticipation and delay play complementary roles. We shall extrapolate on this idea in Chapter 11.

Some of the examples above refer to our nested temporal perception, others to external nested structures. When internal rhythms lock into external ones or vice-versa, we can determine the accompanying boundary shift in terms of the boundary complexity introduced earlier in this chapter.

Sometimes, a boundary arises from an interference pattern created by internal rhythms which overlap with external ones. An example is interference patterns arising from otoacoustic emissions – that is to say, the measurable sounds emerging from the ear. David Kemp suggests that no incoming signal would be perceived unless the receiver also produces a signal – an outgoing one which interferes with incoming air pressure waves. Perception is not passive, but active: what we perceive are the interference patterns of outgoing and incoming waves [20]. Anthony Moore suggests that this would demonstrate how "certain physical aspects of hearing take place outside the body, at the entrance to the ear," thus shifting the boundary between oneself and the world outwards as "an ethereal skin of interference" [21].

Before we look at more abstract types of boundaries in Chapter 5, where we introduce the notion of potential space, a few words on how specific boundaries are implicit in certain models and methods. We all use models of ourselves and the outside world – consciously or unconsciously – and are usually not aware of the fact that we implicitly assume a division between the observer and the observed, namely, a boundary which separates the inside from the outside. Depending on our epistemological approach – that is to say, what we assume we can know at all about the world – our models presuppose very different boundaries. To take two extreme positions, a naive realist believes that there is a real world outside his brain and body, which exists independent of any observer – including himself – interacting with it. A radical constructivist, by contrast, assumes that reality is a construction of his brain and his interaction with the 'non-me' – that is to say, the rest of the world, which includes other individuals and societies.

Fig. 4.2. The view from within: Intentionalism – the purpose and intent inherent in and preceding all human interaction. [22]

Our models also differ greatly when we compare the view from within with an external viewpoint. The view from within corresponds to the phenomenological approach, which does not attempt to explain, but describes the world as experienced from the 1st-person perspective. Phenomenology deals with the world as it appears to us. It makes no claims about the nature of the *noumenon* – the "world behind the phenomena" – and is thus deeply rooted in the Kantian tradition, which is based on the idea that we have no access to the thing-in-itself, but only to things as they appear to us, the *phenomena*. Thus, we have no access to the "real" world – should there be such a thing – but only to perceptions from which we may draw conclusions as to the nature of a reality which exists independent of anyone observing or participating in it. Apart from subjective experience, the phenomenologist is concerned with intentionalism – the purpose and intent inherent in and preceding all human interaction. But phenomenology is not mere introspection – beyond the internal viewpoint, it relates to the external world as a result of being nested into a physical, social and cultural environment.

In contrast to the 1st-person perspective of the endo-observer, a cognitive scientist may take the 3rd-person perspective of a pseudo-external observer, whose studies include explanatory attempts based on psychological motivation, neurological correlations and biological predispositions. I use the expression "pseudo" because no human being can ever be a pure external observer – we are always embedded in social, cultural or linguistic frameworks, so the cognitive scientist is really also an endo-observer. External perspectives are idealizations which are reserved for divinity or philosophical constructs. A famous example of such a construct is Laplace's demon, an ideal observer invented in 1814 by the mathematician and astronomer Pierre Simone de Laplace. This hypothetical entity in the guise of a "pure observer" does not interact with the system under observation (although he is part of that system). As a personification of determinism, he is able to compute the present state and predict all future states of the universe from the past and present position and momentum of all its components. The vantage point of mortal observers, however, is located within the universe, embedded in time and space. Therefore, any attempt to describe temporal structures is limited by *a priori* contextual constraints such as the corrective distortions discussed earlier, which shape our temporal perspectives.

The phenomenological method proves to be a useful candidate for describing the structure of our temporal interface, as it allows us to focus on constraints and corrective distortions inherent in our perspective. This is so because the phenomenological procedure leaves aside ("brackets") our potentially naive worldview which takes the world for granted and forces us, instead, to question the role our perception and consciousness play in the generation of reality. Furthermore, the method of phenomenological reduction (the Latin term *reducere* means "to lead back to" and, in this context, refers to the way the world manifests itself to us) allows us to "adopt the phenomenological attitude, [in which] we are no longer primarily interested in *what* things are – in their weight, size, chemical composition, etc. – but rather in *how* they appear, and thus as correlates of our experience." [23] This approach not only allows for manifestations of our temporal perspectives such as subjective duration but also for pseudo-external points of view and possible interconnections between levels of description. Examples are Douglas Hofstadter's strange loops or tangled hierarchies, which are described later in this chapter.

Earlier, we said that the way we perceive events depends to a large extent on the model we have of ourselves and the outside world. However, when it comes to modelling the world, we encounter very basic difficulties which cannot be circumvented. One major problem was nicely summed up by Norbert Wiener's observation that "the best material model of a cat is another, preferably the same cat" [24]. Because a model never comprises all aspects of a system, we are always forced to deal with an abstraction. As human beings, we have limited time and resources and, as a result of our subjective perspectives, limited access to the world. So when we model a system and make predictions, our pragmatic approach is to focus on one parameter or a few parameters only and to disregard others. To forecast the weather, for instance, we feed a limited number of parameters into a simulation program. We may include parameters such as temperature, air pressure and prevailing winds but will stop short of including every motion on the planet, such as the flapping of a butterfly's wings. We proceed in this way for practical reasons, although we know that the butterfly effect may well be that decisive tiny impact which tips the weather system to develop into full-blown hurricane or a calm and sunny day. In this book, we are primarily interested in internal models or, in Dubois' terms, the property of strong anticipation. This property determines human behaviour and is found in all structures or processes which contain a model of themselves and their environment and refine those models by interacting with their environment.

Depending on our choice of methods and models, the boundaries which define inside and outside are set at very different locations. Let us take a brief look at three cuts, which all connect and separate inside and outside, but rest on very different worldviews: The Cartesian Cut, the Heisenberg Cut and the Rössler Cut. Next, we proceed, via Einstein's relativistic observer frames, to Otto Rössler's endophysics and Roger Boscovich's covariance principle. Finally, we shall take a brief excursion into Douglas Hofstadter's tangled hierarchies in order to work out differences between internal and external temporal perspectives.

For our purposes – that is to say, to describe temporal perspectives – the following descriptive features shall suffice to differentiate between cuts, interfaces and boundaries. A cut creates separate worlds and is therefore a philosophical division. An example is Cartesian dualism, which divided human beings into two separate realms: the mental and the physical. An interface, on the other hand, is an interactive filter which connects systems. These may be as diverse as social organizations, computers, life worlds, boundary surfaces in chemistry, biofilms or software. I define a boundary as an extended field which connects and separates systemic wholes and may consist of a number of nested interfaces.

The first cut we consider here is a division which has had a substantial impact on Western science: the Cartesian cut. In the seventeenth century, René Descartes distinguished things physical from things mental (*res extensa* and *res cogitans*) and thereby created the mind-body problem, which deals with the question as to how the two relate to each other [25]. Although Descartes' division has shaped our scientific theories for centuries, recent advances in cognition and embodiment suggest that the Cartesian Cut is not a helpful model [26]. The mind is embodied – that is to say, our cognitive performances are constrained by our limited perceptual apparatus and by metaphors which are directly linked to our sensory-motor system. George Lakoff and Rafael Núñez' have described some of these constraints in their seminal work on embodied mathematics [27]. A very simple example is embodied arithmetic, which is based on the idea that there is a precise mapping from the domain of physical objects to the domain of numbers. Examples of such mapping include "addition as adding objects to a collection, subtraction as taking objects away from a collection, sets as containers, members of sets as objects in a container." [28] Another example is how counting with our ten fingers invites us to use grouping, ordering and memory capacity which creatures without fingers would very probably have not evolved. Lakoff and Núñez suggest that

> such regular correlations (...) result in neural connections between sensory-motor physical operations like taking away objects from a collection and arithmetic operations like the subtraction of one number from another. Such neural connections, we believe, constitute a conceptual metaphor at the neural level – in this case, the metaphor that Arithmetic Is Object Collection. [29]

As our models are based on abstractions constrained by our sensory-motor system and the resulting conceptual metaphors, the Cartesian Cut between *res cogitans* and *res extensa* is, in the light of the embodied mind, not a useful model to describe our temporal perspectives.

Einstein's interfacial cut, which came in the shape of observer frames whose relative position and motion determine the successive order of events and measured simultaneity, is a necessary but not sufficient ingredient for modelling a temporal perspective. It is not sufficient because the relativistic notion of a temporal observer frame does not take account of either the interference any act of observation implies on the micro-level nor of the observer's internal microscopic differentiation. The first issue was encountered in quantum theory, which set the Heisenberg Cut and the second in endophysics, introducing the Rössler Cut.

At first sight, the Heisenberg Cut shifts the line of division between the observer and the observed towards the outside. It seems to be no longer a division between mental and physical things as performed by the Cartesian Cut, but rather a separation of the observer and the observed, with the measuring chain usually attributed to the system "observer". However, on closer inspection, the Heisenberg Cut is not completely disjoint from Descartes' cut between the mental and the physical world. The outcome of the measurement depends, after all, on the decision as to where exactly the cut is located. And that decision is usually made by a flesh-and-blood human being with working cognitive functions [30]. It is by no means a trivial matter to decide where the measuring chain ends and the object to be observed begins. We can see the difficulty already in the comparatively transparent example of the unilateral neglect patient who used a stick for reaching and thereby virtually doubled his event-horizon. He was able to perceive and control the so far neglected hemisphere because the use of the tool led to a remapping in his brain of far space to near space.

So far, we have pretended that something like an "observer" exists, who can scrutinize a system without influencing it. But quantum theory shows that measuring means interfering: Whether we probe a system by shooting light at it or electrons as we do in microscopy or sound waves in a sonographic measurement, the principle is the same. We throw something in the direction of the phenomenon we want to measure and draw conclusions about the nature of that phenomenon by interpreting the interference patterns produced by our disturbance and the system's reaction to that disturbance. Any act of observation at that level is, in fact, an interference, just as a measurement is. So there are no pure observers at the quantum level, but only "observer-participants", since in every observation, the observer is also an actor – that is to say, he participates in the system. So this may be a convenient point at which to abandon the notion of a pure observer for all practical purposes and, henceforth, talk of observer-participants.

Observer-participancy blurs the separation between inside and outside, mind and matter, subject and object [31]. On the macro-level, this dichotomy is blurred only if we take account of the observer-participant's microscopic movements. Enter Rössler's micro-relativity, which separates the inside from the outside – that is to say, the observer-participant from the rest of the world. The Rössler Cut divides the world into endo and exo – the view from within and the view from the outside. This new cut originated from Rössler's endophysics, which describes physics from within, that is a physics which includes the observer-participant – in other words, the microscopic movements in his current physical state – and is described from that observer-participant's perspective. A person who is running a fever perceives the world differently from someone whose body temperature is normal. One result of endophysics is the realization that we are all endo-observer-participants, embedded in the system we describe. There is no way we could ever take on an external vantage point and look at the world from the outside. Again, such an exo-perspective is reserved for a super-observer, a divinity or Laplace's demon. All the endo-observer-participant may talk about is the world as it appears on his interface with the rest of the world – his Now – which would produce a purely phenomenological account.

In a nutshell, Rössler's endophysical approach – which he also refers to as microconstructivism or micro-relativity – states that the embedded observer-participant is always limited to an endo-perspective. The difference between the endo- and exo-perspective reveals itself in the fact that our Nows manifest as very private event horizons, generated by the microscopic movements within the observer. And the Now is our only access to the world: What we see is *our* interface reality.

However, this conclusion does not necessarily entail that communication is pointless or impossible. As the Austrian-British philosopher Karl Popper pointed out, however restricted people may feel they and others are by their own private, social, and cultural reference frames, on a pragmatic level, it still makes sense to look at the logical *consequences* of theories – or frameworks, as Popper denotes them – and find out which are preferable to us. He suggests that, even in the presence of a seemingly unbridgeable gulf, we can learn about the structure of our own constraints from the mere fact that a clash of frameworks has taken place. In fact, it is only because of this clash that we become aware of the fact that there are things like personal cultural, moral or logical frameworks which constrain our own perception of the world. According to Popper,

(...) frameworks, like languages, may be barriers. They may even be prisons. But a strange conceptual framework, just like a foreign language, is no absolute barrier: we can break into it, just as we can break out of our own framework, our own prison. [32]

The more clashes we experience, the more aware we become of the structure of our own frameworks. Analogously, the more an observer-participant is aware of his physical and cognitive constraints which lead to the corrective distortions generated by his internal structure and dynamics, the more steerable becomes his Now, his temporal interface.

It is true that the kind of invisibility which accompanies the merging of interfaces, as in the examples of the walker with the stick, the macaque monkey and the neglect patient, enables us to navigate the world more smoothly. But it also prevents us from becoming aware of the interfacial distortions which determine temporal and spatial order and shape our perspective. In our everyday lives, our potential awareness of the microscopic processes within the observer, as well as his cognitive structures, is limited. We cannot monitor our thinking on low-level thought processes, so most of it remains inaccessible to conscious introspection [33]. But, as Thomas Metzinger has pointed out, naive realism has proven to be an extremely useful perspective for navigating and dealing with the world and is thus a selection effect:

(...) naive realism has been a functionally adequate assumption for us, as we only needed to represent the fact 'there's a wolf there' (...) not 'there's an active wolf representation in my brain now'. [34]

The Rössler Cut differs from the Cartesian and Heisenberg Cuts in that it is not simply a clear division between the material and the mental world or between the observer-participant and the observed. The interface between inside and outside includes the human observer-participant [35] – we may say that it is necessarily an extended boundary. This boundary shift towards the inside means that we have to take into account the microscopic movement within the observer when we model reality [36].

As Rössler has pointed out, there is a 250-year old principle which preceded both Einstein's principle of relativity and Rössler's endophysics. It expresses the idea that an observer cannot distinguish whether a change he experiences is the result of a change in the external world or within himself: The Boscovich covariance principle, named after Roger Joseph Boscovich, a Jesuit astronomer, physicist and philosopher, states that the world can only be described relative to an observer. "(...) a state of external motion of the observer relative to the world is *equivalent* to a state of motion of the whole world relative to a stationary observer" [37]. Of course, to us and our contemporaries, who have been taught Einstein's Theories of Relativity at school, this does not come as a complete surprise. However, there is more to Boscovich covariance than mere relativity of motion. It also applies to the state of the observer's internal motion in relation to the rest of the world and claims that the observer cannot observe the world as it is, but only the interface between him and the world [38]. In other words, if the microscopic movements within the observer and those of the world around him were time-inverted, the observer would detect no change. This is so because the two changes would compensate each other and level out, so to speak, the difference which makes up the interface – which is all the observer can access.

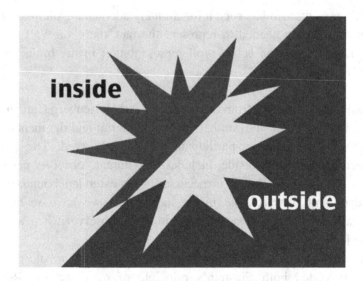

Fig. 4.3. Inside/outside inversion on our interfaces? [39]

If we were to detect a time-inversion, this would mean that a time-reversal had taken place *either* within us *or* in the external world. Which of these we would be unable to decide, as we could see only the time inversion of our interface. And just as we cannot rule out that there are hidden correctives performed unconsciously by the observer, as we have seen in Chapter 3, we must also accept that the existence of compensatory changes inside and outside the observer cannot be excluded *a priori*.

> If only the observer's internal motions are time inverted, this is equivalent to no change having taken place in the observer, but rather to the external world having been time-inverted instead. This means that any change inside the observer that can in principle be exactly compensated for by some external change in the environment (so that the net effect on the interface would be zero), is equivalent to that compensatory change having occurred objectively in the environment with no change having taken place in the observer. [40]

What follows in both cases is Kant's sobering insight that we can never see the world objectively or, in the parlance of endophysics, take the viewpoint of an external observer. What we can do, however, is to question the assumptions implicit in our endo-perspectives. To reveal corrective distortions is a first step.

Another way of uncovering hidden structures is to look for higher hierarchies – that is to say, embedding LODs whose existence we can only make out indirectly. This can happen when we encounter a type of logical inconsistency which Douglas Hofstadter denoted *strange loops* or *tangled hierarchies* (both terms are interchangeable). [41] An example of such a tangled hierarchy is the authorship triangle: Author A exists only in a novel by B. Likewise, B exists only in a novel by C. And C exists only in a novel by A. There is, of course, a way out of this tangle, namely, if A, B and C are all fictitious characters in a novel by D. While D can influence A, B and C, the tangled authors cannot touch author D's

LOD – in fact, the only way they can guess that such an embedding level exists is by recognizing the strange loop on their own LOD.

According to Hofstadter, beyond every strange loop, there is an inviolate level, which cannot be influenced by lower levels. For instance, A, B and C cannot influence D's actions. A famous example of a strange loop is Escher's famous etching Drawing Hands (see Figure 4.4 for a replica). If we look at the action of writing or drawing, each is an example not only of a logical tangle but also of a temporal strange loop: Does the left hand first draw the right one or vice-versa? Which writer started the authorship triangle? Of course, the strange loop only exists if we are limited to one LOD. As soon as we take a step back, we can make out an embedding LOD, which cannot be influenced by the nested level. In the case of the authorship triangle, it is the writer and inventor of A, B and C, who forms this higher, embedding level, and in the etching, it is Escher who draws the two hands from a higher LOD.

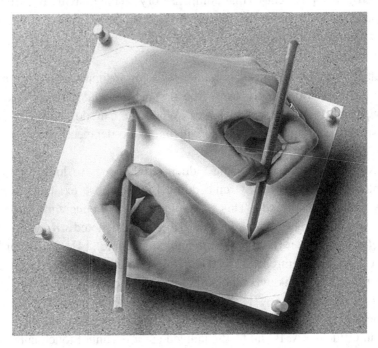

Fig. 4.4. A strange loop or tangled hierarchy: "Homage to Escher's Drawing Hands". [42]

If Hofstadter is right and there is an inviolate level beyond every tangled hierarchy, there is always a part of the observer-participant which does not participate. A strong case can be made for the existence of an inviolate level, but the boundary between the non-participating core and the participating part of an agent seems to be highly negotiable, to say the least [43]. But, as mentioned earlier, a pure observer – that is to say, someone who does not interact with his context – is inconceivable because he could not let his environment either know or feel the consequences of his observations. (If he did, he would have to interact in some way.) A pure participant, by contrast, would be in a state of total immersion. More than this, he would form a whole with Nature and, in the spirit of Zen, be at one with the universe.

However, Hofstadter's distinction between self-modifiable software and inviolate hardware is not as clearcut as it seems at first sight. Sometimes experimenting leads to learning which, in turn, leads to changes on the neural level, as we saw in the case of the macaque who adjusted to the delay: conditioning blurs the boundary between self-modifiable temporal interfaces and their (potentially inviolate) temporal context.

Although nested contexts can be pictured as concentrically nested, which resembles the inside-outside relation as portrayed by endo- and exo-perspectives, they are usually pictured as vertical hierarchies. And the topology of vertical hierarchy brings forth causal relations in the directions of increasing and decreasing complexity. I am referring here to the notions of *upward* and *downward causation*. The German theoretical physicist Hermann Haken has looked at how patterns form and are maintained through the interaction of the whole and its parts – a science he calls synergetics [44]. Synergy is the simultaneous causation of upward and downward directions:

> The upward direction is the local-to-global causation, through which novel dynamics emerge. The downward direction is the global-to-local determination, whereby a global order parameter 'enslaves' the constituents and effectively governs local interactions. There is no supervisor or agent that causes order; the

system is self-organized. The spooky thing here, of course, is that while the parts do cause the behaviour of the whole, the behaviour of the whole also constrains the behaviour of its parts according to a majority rule; it is a case of circular causation. [45]

Circular causality seems to be a feature of self-organizing systems – that is to say, systems which display emergent behaviour (i.e., where the whole is more than the sum of its parts). Haken's original example is the laser, where the individual photons organize in a structure and are, at the same time, organized by that structure – that is to say, the order parameter. In this case, the cause is not to be found in either the global or the local action but is the result of the interaction of both. But although this may be reminiscent of Hofstadter's strange loops, there is an important difference between the latter and Haken's synergetics. Whereas strange loops, which occur on one LOD, can be disentangled by contextualization – that is to say, by taking a step back – circular causality in synergetics operates *between* LODs and cannot be disentangled by contextualizing, i.e., by nesting the system into an embedding LOD. Circular causality in self-organizing systems is not a conjuring trick whose smokescreen can be wiped away by adding context. It is inherent in such emerging systems and a defining feature of self-organization. A mere boundary shift will not reveal an explanation.

In Chapter 2, I briefly referred to Hofstadter's remark that we must distinguish between what we generally denote as physical causality and the description of a phenomenon on a different LOD, which is often a mere translation or explanation rather than a material or efficient cause (*causa materialis* or *causa efficiens* in Aristotle's terms). Menno Hulswit makes the same point when he suggests that 'downward causation' is a misleading term:

(We) have concluded that 'causation' in respect of 'downward causation' is usually understood in terms of explanation and determination rather than in terms of causation in the sense of 'bringing about'. Thus, the term 'downward causation' is badly chosen. [46]

Whenever self-referential systems create circular causation on one LOD, a meta-level usually resolves the tangle. However, this can prove to be very confusing when the meta-level adheres to the same rules, for instance, grammar rules. For that reason, language creates strange loops when it refers to itself:

> Here, something in the system jumps out of the system and acts on the system as if it were outside the system. What bothers us is perhaps an ill-defined sense of topological wrongness: the inside-outside distinction is being blurred, as in the famous shape called a 'Klein bottle'. [47]

Fig. 4.5. The Klein bottle: blurred boundaries between inside and outside. [48]

The idea of blurred boundaries between inside and outside, or subject and object, is not a new one and has been taken to extremes in the philosophy of Zen Buddhism. Zen distinguishes three stages of the externalization of the internal. At the first stage, there is a clear separation of inside and outside, of self and non-self:

> The initial stage, corresponding to the world-experience of an ordinary man, at which the knower and the known are sharply distinguished from one another as two separate entities, and at which a mountain, for example, is seen by the perceiving 'I' as an objective thing called 'mountain'. [49]

The second stage is that of absolute unification, which precedes any separation of subject or object:

> At this stage, the so-called 'external' world is deprived of its ontological solidity. Here the very expression: 'I can see a mountain' is strictly a false statement for there is neither the 'I' which sees nor the mountain which is seen. [50]

At the third stage, the undivided oneness divides itself into self and non-self. However, despite the apparent division, oneness is kept intact – that is to say, subject and object are both separated and merged into each other.

The opposite process – that is to say, the internalization of the external – is a process by which Nature becomes established as an 'internal landscape'. Interestingly, as Toshihiko Izutsu points out, the underlying spiritual event of both – the internalization of the external and the externalization of the internal – are the same [51].

We have described boundaries as extended fields which consist of a number of nested interfaces, some of which are invisible to the endo-observer but detectable from the exo-perspective. As a result of this difference, it has been possible for a boundary complexity to be defined. In the next chapter, we shall continue to look at how the internalization of the external and the externalization of the internal shape the structure of the boundary between self and non-self. Against this background, special focus will be given to the notion of temporal contextualization.

References

1. A. Clark, *Supersizing the Mind. Embodiment, Action and Cognitive Extension.* Oxford University Press, New York 2008, pp. 30-37.
2. Wikipedia entry on "systems theory", December 2009.
3. A. Clark, pp. 30-37.
4. Photograph courtesy of Manuel Atienzar 2009.
5. V. S. Ramachandran and S. Blakeslee, *Phantoms of the Brain.* Fourth Estate, London, 1999.
6. A. Clark, pp. 30-37.
7. M. Botvinick and J. Cohen, Rubber Hands 'Feel' Touch That Eyes See. In: *Nature*, Vol. 391, 1998, p. 756.
8. G. Buzsáki, *Rhythms of the Brain.* Oxford University Press, New York 2006, p. 6.
9. ibid pp. 5-6.
10. O. van Nieuwenhuijze, *Integral Health Care.* Dissertation draft, 1998 (unpublished) and personal communication 2005.
11. K. W. Kratky, Chronobiology and Cross-Cultural Medicine: Cyclic Processes during a Day, a Year an a Lifetime. In: *Health, Healing and Medicine – Vol. XI. Comparative and Integrative Medicine.* Ed. by George E. Lasker and Karl W. Kratky. Proceedings of the 16th International Conference on Systems Research, Informatics and Cybernetics, Baden-Baden 2004.
12. A. Clark, pp. 30-37.
13. O. E. Rössler, *Endophysics.* World Scientific, Singapore 1998.
14. S. Vrobel, Negotiating Boundaries. In: *Cybernetics and Systems.* EMSCR Conference Proceedings, Vienna 2010.
15. C. Stetson, X. Cui, P. R. Montague, D. M. Eagleman, Illusory reversal of action and effect. *J. Vis.* 5: 769a, 2005.
16. H. Bergson, *Einführung in die Metaphysik.* Diederichs, Jena, 1909, p. 8, my translation.
17. ibid p. 27 ff.
18. D. M. Dubois, New Trends in Computing Anticipatory Systems: Emergence of Artificial Conscious Intelligence with Machine Learning Natural Language. In: *Computing Anticipatory Systems*, AIP Conference Proceedings Vol. 1051. Ed. by D. M. Dubois, New York 2008, p. 30.
19. ibid p. 26.
20. D. T. Kemp, Otoacoustic Emissions. In: *Some Perspectives*; Theodore J. Glattke, University of Arizona, US, 3/4/2002.
21. A. Moore, Transactional Fluctuations I. In: *Variantology 3; On Deep Time Relations of Arts, Sciences and Technologies*; Ed. by Siegfried Zielinski & Eckhard Fürlus. Verlag der Buchhandlung Walther König, Cologne 2007 and personal communication 2009.

22. Photograph "Inside/Outside" courtesy of Maria José Pablo 2009.
23. S. Gallagher and D. Zahavi, *The Phenomenological Mind – an introduction to philosophy of mind and cognitive science.* Routledge, New York 2008, p. 25.
24. A. Rosenblueth and N. Wiener, The Role of Models in Science. In: *Philosophy of Science* 12, 1945, p. 316.
25. R. Pfeifer and J. Bongard, *How the Body Shapes the Way we Think – a New View of Intelligence.* MIT Press, Cambridge, 2007, p. 6.
26. Cf.: e.g. M. Storch et al, *Embodiment. Die Wechselwirkung von Körper und Psyche verstehen und nutzen.* Huber, Bern 2006.
27. G. Lakoff and R. E. Núñez, *Where Mathematics Comes From: How the Embodied Mind Brings Mathematics into Being.* Basic Books, New York 2000.
28. ibid p. 53.
29. ibid p. 55.
30. R. Pfeifer and J. Bongard, *How the Body Shapes the Way we Think – a New View of Intelligence.* MIT Press, Cambridge, 2007.
31. *Inside Versus Outside. Endo- and Exo-Concepts of Observation and Knowledge in Physics, Philosophy and Cognitive Science.* Ed. by Harald Atmanspacher and Gerhard J Dalenoort. Springer, New York 1994, p. 9.
32. K. R. Popper, *The Myth of the Framework.* Ed. by M. A. Notturno, Routledge, New York 1994, p. 61.
33. Metzinger suggests that this inaccessibility is a result of the fact that the perspective-generating mechanisms are transparent to us because they occur on time scales which are far too fast for us to be accessed consciously. See: T. Metzinger, *Being No One: Consciousness, the Phenomenal Self and the First-Person Perspective.* Foerster Lectures on the Immortality of the Soul. The UC Berkeley Graduate Council, 2006.
   http://video.google.com/videoplay?docid=-3658963188758918426&q=neuroscience
   and personal correspondence 2006.
34. ibid.
35. *Inside Versus Outside. Endo- and Exo-Concepts of Observation and Knowledge in Physics, Philosophy and Cognitive Science.* Ed. by Harald Atmanspacher and Gerhard J Dalenoort. Springer, New York 1994, p. 7.
36. O. E. Rössler, 1998, and personal communication 2007.
37. O. E. Rössler, Boscovich Covariance. In: *Beyond Belief. Randomness, Prediction and Explanation in Science.* Ed. by John L. Casti and Anders Karlquist CRC Press, Boca Raton 2000, p. 66.
38. ibid p. 66.
39. Illustration courtesy of Matthew Bell 2009.
40. O. E. Rössler, 2000, p. 67.
41. D. R. Hofstadter, *Gödel, Escher, Bach: An Eternal Golden Braid.* Penguin 1980, p. 686.
42. Illustration courtesy of Tim Kondermann 2009.

43. S. Vrobel, The Kairos Syndrome. In: *International Journal of Computing Anticipatory Systems.* Vol. 20, Ed. by D. M. Dubois, CHAOS, Liège 2008.

44. H. Haken, *Erfolgsgeheimnisse der Natur. Synergetik: Die Lehre vom Zusammenwirken.* Rowohlt, Reinbek 1995.

45. G. Buzsáki, 2006, p. 14.

46. M. Hulswit, How Causal is Downward Causation? In: *Journal for General Philosophy of Science.* Vol. 36, No. 2, Springer, Netherlands 2005.

47. D. R. Hofstadter, 1980, p. 691.

48. Illustration courtesy of Hua-Chun Chao 2010.

49. T. Izutsu, The Interior and Exterior in Zen Buddhism. In: *Eranus Jahrbuch 1973: Correspondences in Man and World.* Ed. by A. Portmann and R. Ritsema. E. J. Brill-Leiden, Netherlands 1975, p. 608.

50. ibid p. 608.

51. ibid p. 609.

Chapter 5

Contextualization:
Embedded Observer-Participants

There was little agreement among ancient Greek philosophers about how we see external objects. Opinions were divided on whether and how we internalize the exterior or externalize the interior. The atomists believed in the intromission theory – that is to say that, for an object to be visible, its particles must enter the eye. Be it as atoms flowing in a continuous stream from the object into the eye, as Epicurus suggested, or as pressed air, which holds the object's colours, as proposed by Democritus. Euclid, by contrast, believed in the extramission theory, which holds that visual rays emanate from the eye in the shape of a cone (see Fig. 5.1).

Fig. 5.1. Extramission: Visual rays emanating from the eye. [1]

The Platonic tradition defended a combination of intromission and extramission, in which a stream of fire or light, which is emitted from the eye, fuses with the light of the sun. Aristotle accepted neither theory and argued that, if rays emanate from the eye, it should be possible to see in the dark. And the idea that objects shrink and enter the eye also seemed absurd to him. He came up with the idea that sunlight is reflected by objects and transmitted into the eye through a medium [2]. With a few notable exceptions, this is roughly what most people believe today. A modern advocate of the theory that eyesight results from a combination of intromission and extramission is Max Velmans, who maintains that that when we perceive an object via light rays through the eye, it is turned into a neural representation by our brains and then projected outwards, thus creating phenomenal space [3].

The question as to whether perception is active, passive or a combination of both is not limited to optical considerations. Rössler has posed the question of how we create the wallpaper in our brains [4], i.e. how do we generate a 1st-person perspective and a shared reality? We certainly do not just map the outside world like a *camera obscura*. Nor are we monad-like closed systems with, as Leibniz put it, no windows to the outside world [5]. We saw in Chapter 4 that the world-observer boundary can expand and contract, depending on whether we assign parts of the environment to ourselves or vice-versa. These assignment conditions are decisive when it comes to our perception of the world because the resulting extended boundary forms our personal reality.

The interplay of emissions both from the outside and from within us creates interference patterns, which form the wallpaper in our brains. This happens, for instance, when oto-acoustic emissions interfere with external sound waves. (Oto-acoustic emissions are sounds created by our inner ear. The cochlear, which amplifies external sounds, increases its activity in the absence of external acoustic stimuli and thus generates oto-acoustic emissions.)

In general, intromission is everything that makes an impression on us from the outside, in the sense that it modifies our interfaces by generating "in-formation" – that is to say, by making "indentations". Such an impression can be limited to one sense, for instance, when we listen to music inside an isolation tank (a tank which is dark, soundproof, half-filled with body-temperature water and used to simulate sensory deprivation). Usually, however, we perceive the world around us via several senses simultaneously, which means that what we perceive is an interference pattern. An example of such cross-modal interference – that is to say, two or more senses influencing with each other, is speech perception. We detect sound waves not only through the ear but also via receptors in our skin, which register tiny bursts of aspiration produced by some speech sounds such as 'pa' or 'ta'. Gick and Derrick showed that tactile information is integrated into auditory perception in much the same way as has been observed in the coupling of auditory-visual stimuli [6]. In the experiment, participants were exposed to inaudible air puffs on the hand and the neck and, simultaneously, to aspirated speech sounds. Subjects tended to hear all syllables co-ordinated with those air puffs as aspirated. For instance, they would mishear 'b' (hardly aspirated) as 'p' (aspirated) if the sound was accompanied by a cutaneous air puff: the interference of external cross-modal stimuli sufficed to distort the interpretation of the overall impression. Perceptual distortions result from interference among stimuli which reach us via our various sensory channels as well as interference resulting from the interplay of internal and external dynamics. Those internal dynamics can cause extramissive interference patterns either by emitting, say, electromagnetic waves, when our epidermis dissipates heat, or by a specific mindset which acts as a cognitive and perceptual filter. We shall discuss examples of cognitive filters which turn into perceptual ones later in the chapter.

In a nutshell, the extramission which leads to such interference is manifested in what Weibel calls "the noise of the observer" [7]. Following Rössler's observation that the world as it appears on our interfaces is always coloured by the observer's internal microscopic movements, Weibel states that the observer's own signal or noise and that of his object become inextricably intertwined. A dilemma arises if the observer is not aware that he is an endo-observer and misinterprets his own noise as information about the situation under observation. This would happen in the case of total sensory deprivation. Weibel goes as far as saying that it is the noise of the observer which manufactures cultural consent, be it in science or religion. The point is that, as observers, we

cannot rid ourselves of the "interference" our own internal noise causes –
we *are part of* the interference pattern:

> The noise of classical communication theory is more or less the
> noise of one's own signal, where the observer acts to correct
> errors. The noise in quantum physics is the noise of the observer,
> unavoidably and necessarily producing errors. (...) Information is
> therefore unavoidably observer-relative. Of necessity the observer
> creates noise. He can escape this noise of observation only by
> himself becoming a part of the information model. [8]

A way of circumventing self-generated noise would be to take a step
back and look at the system from the outside, as a super-observer. In
practice, this is not achievable as an uninvolved observer, such as
Laplace's Demon, does not exist. Hofstadter referred to this kind of
level-crossing, in which internal and external perspectives intertwine, as
"strange loops" or "tangled hierarchies" (see Chapter 3). A strange loop
occurs, when something within the system jumps out of it and acts upon
that system as if it were outside it [9]. By definition, a strange loop can
only be resolved by taking a step back onto a meta-level – that is to say,
by adding a dimension or introducing another degree of freedom into the
system – for instance, an additional observer who is situated outside the
tangled hierarchy.

Having accepted that, by generating internal noise, we are part of the
interference pattern, we must recognize that it is of minor importance
whether the observer-generated noise is actually emitted into our close
environment (i.e. running a temperature and dissipating body heat) or
remains below our epidermis (e.g. experiencing hunger). As long as there
is internal structural change, our interface reality will be affected by it
just as much as by external impacts. This is a variation on Boscovich's
co-variance effect we discussed in Chapter 4. And, although, as Rössler
has pointed out, too much symmetry should make us suspicious [10], we
are rarely aware of co-variance, as the structure of our interfacial filters
has become transparent to us. Weibel is aware of our blindness when it
comes to recognizing how co-variance shapes our perspective:

Observation by an observer is, therefore, no longer sufficient to increase information; rather, what is required is an increased correlation and co-variance of observers and observations. It is questionable, however, whether we can grasp these correlations. [11]

This blindness is probably a selection effect. It is also often a blessing in disguise, since it saves time and energy to avoid conscious filtering, as Metzinger's wolf example shows. It is sufficient, he says, to represent the fact "there's a wolf there" without having to add another bracket by nesting our percept into the phenomenological interpretation: "there's an active wolf representation in my brain now" [12]. It is only when our expectations are not met or misattributions occur, that it becomes useful to make visible the usually transparent internal structure of our perspective.

The two-way interaction between the observer and the rest of the world generates the interference pattern we discern as reality. As social beings, most of us constantly seek out interaction with our environment, which includes other people. This interaction continuously modifies our world-self boundary and creates what Moore refers to as an ethereal skin of interference (ESI), a simultaneous interaction between extramission and intromission [13].

Other impacts on the interference pattern which manifests itself as our reality are cognitive and perceptual filters induced by conditioned responses. And as our actions are goal-directed, these show themselves as (among other things) intentions, expectations and attentional focus. Attention, for instance, focusses and constrains our perspective and is a compelling example of how a cognitive filter can give rise to a perceptual one. To see how this reveals itself, let us look at the phenomenon of change blindness. This occurs when someone fails to notice (sometimes very large) changes to objects or scenes in his environment. This failure can happen from one instance to the next. Simons and Levin have shown that individuals who had been asked for directions by a pedestrian in the street often did not notice that that pedestrian was replaced by another while a door was carried in front of the first one [14].

With change blindness, failure to detect modifications from one moment to the next could be memory-related, as successive impressions need to be connected. To avoid a possible bias caused by memory performance, another phenomenon, called attentional blindness, is often presented as an even stronger constraint on our perception. In attentional blindness, all the information is potentially available simultaneously (not successively, as in change blindness), provided we pay attention. It reveals that an event which happens in our environment is not perceived at all if we do not focus on it – that is to say, if we do not give it our attention. A famous example is the gorilla-like being (actually a woman in a gorilla costume), who walked through a basketball game, unnoticed by approximately half of the test subjects who watched a recording of it. As the game absorbed all their attention, anything unexpected became invisible to them. The experiment was repeated later in a slightly varied form, with the image of a woman with an umbrella being superimposed on the screen. Here, too, approximately half of the test group failed to notice the woman. Simons and Chabris observed in their experiments that the duration of an unexpected event that can be missed is surprisingly long: In the first experiment, the test group were shown a stimulus tape of 63-seconds, in 9 of which the gorilla was visible, walking from right to left into the basketball court, stopping in the middle and looking into the camera, thumping its chest and continuing to walk across the court [15]. As we saw in the first chapter, a narrowed focus can blot out not only visual stimuli, but any modality. The police officer failed to hear his colleague's gunshots and found his field of vision greatly reduced because he focussed on his potential target. But, as the gorilla experiment shows, it does not need a stressful situation like a shoot-out to constrain our perceptions. We often perceive only those objects and details which receive focussed attention, while unanticipated objects and events can go unnoticed.

Another cognitive filter which gives rise to perceptual constraints and distortions is what Ellen Langer calls *mindlessness* – a mental state in which we rely too rigidly on categories created in the past. Langer calls this condition "entrapment by category" [16]: Imagine someone ringing your bell in the middle of the night and offering you ten thousand dollars for a seven-by-three foot piece of wood. You are standing in the open doorway wrecking your brain to think from where you could obtain such a piece of wood quickly. It seems an impossible task, so you decline the offer reluctantly. Next day, passing a construction site, you see a seven-by-three foot piece of wood – a door. You could have just unhinged your door and exchanged it for a small fortune. But your mindlessness hid the piece of wood from you because, to you, it was stuck in the category 'door' [17]. The many causes of mindlessness range from automatic behaviour, a belief that resources are limited, acting from a single perspective and the constraints of linear time to the influence of context. For our purposes, the last two causes are of particular interest.

Let us take a closer look at the notion of context. The term is derived from the Latin *contextus*, meaning 'joining together' (from *com* = together, *textere* = weave). The verb *contextere* can be translated as 'weave together' – a sense relevant to the notion of context in temporal perspectives. Here, the notion implies that something new emerges from the 'interference' - namely, the simultaneous and successive weaving together, of several strands of string. These could be internal and external strings woven together into an extended boundary (for instance, oto-acoustic emissions and external sounds). The emergent structure could also result from internal interference, as in cross-modal perception, when our brain moulds visual and auditory stimuli into one event even if they arrive at different times, as in the perception of close-by thunder and lightning). Alternatively, external stimuli may interfere before they interact with the observer, as do superimposed electromagnetic or sound waves. For instance, mobile phone headsets can cancel out unwanted background noise in planes or trains before it hits the ear of the speaker, making it unnecessary for him to raise his voice [18]. However, through his oto-acoustic emissions, the speaker does still interfere with the external interference pattern of waves which cancel out or re-enforce each other. The pattern which emerges is more than the sum of its parts because the individual simultaneous strings which are woven into a new fabric create Δt_{depth} and thus a temporal perspective. And if we look at the way rhythms cause temporal patterns, whether by phase-locking (e.g. clocks on a shelf which fall into a uniform ticking rhythm) or more complex temporal interference patterns (e.g. women shifting each others' menstrual cycles), simultaneity is, in each case, a sufficient condition for the existence of context. In connection with temporal perspectives, I shall henceforth use the term 'context' to mean simultaneous structures or patterns (including people) which are arranged simultaneously in a temporal perspective – in other words, in Δt_{depth}.

In this sense, a lack of context – simultaneous structures or patterns – acts as a cognitive and perceptual constraint. Lack of Δt_{depth} means a lack of complexity, which may result in a pathological condition. Bruce West has shown that disease is a loss of complexity [19]. This loss can be self-inflicted (for instance, by drug abuse) or the result of environmental or genetic effects. In Chapter 9, we shall look into the impact of internal and external complexity on health.

An example of self-imposed limitation is the one-dimensional, one-directional linear notion of time which still largely prevails in Western societies. This limitation – mindlessness – prevents us from perceiving the richness and full beauty of complex systems. One point we thus miss is that simultaneous levels of description often appear to exert both bottom-up and top-down causality. However, this concept is not as clearcut as it may appear. The problem with downward causation is that it means very different things to different people. Davies distinguishes between whole-part causation and level entanglement (the second of which we discussed in connection with Hofstadter's strange loops) [20]. In whole-part causation, the behaviour of a part can be understood only by referring to the dynamics of the whole:

> (...) a ball rolling down a hill implies that each of the ball's atoms is accelerated according to the state of the ball as a whole. But it would be an abuse of language to say that the rotating ball *caused* a specific atom to move the way it did; after all, the ball *is* the sum of its atoms. What makes the concept 'ball' meaningful in this case is the existence of (non-local) constraints that lock the many degrees of freedom together, so the atoms of the ball move as a coherent whole and not independently. But the forces that implement these constraints are themselves local fields, so in this case whole-part causation is effectively trivial in nature. [21]

Davies also observes that even in self-organized systems such as lasers, where atomic oscillators (photons) phase-lock into a coherent light beam, the dynamics can be accounted for fully in terms of local interactions and non-local constraints.

In downward causation as level entanglement, by contrast, higher conceptual levels have a causal impact on lower ones. As Hofstadter has pointed out, in such cases, 'explanation on a different LOD' would be a better term than 'causality' [22]. All mind-brain interaction falls into this category. When I say that I moved my arm and gripped the handle of the coffee cup, volition is interpreted as exercising a causal effect over physical matter.

Both kinds of mechanism work, regardless of whether we wish to call it causation. The second kind – downward causation – is found in complex systems which self-organize, such as human beings. Human action is goal-oriented – that is to say, the purpose of an action must be taken into account when we try to identify the cause. In such cases, final causation is a powerful driving force, therefore, a one-dimensional concept of time is a constraint, since it does not consider simultaneity as an additional degree of freedom (which would be necessary to describe interacting levels which provide context for each other). It is easier to describe temporal aspects of complex systems such as humans in terms of fractal time, because this does not neglect the role of context. Temporal context manifests itself as the dimension of simultaneity, which, together with succession, suffices to describe embedded systems (for instance, an individual who is nested in a family, which, in turn, is nested in a society, which, again, is nested into mankind as a whole). By taking both temporal extensions into account, we escape being trapped in the category of one-dimensionality.

Just as our experience of seeing, hearing and touching results from a combination of passive and active components, our perception of duration depends not only on external time spans but also our expectations and internal states. These internal states can be moods or temperaments, depending on whether they are short-lived whims or attitudes which have been acquired over a long time span. In this connection, it is rewarding to take a closer look at the notion of *embodiment* – the idea that cognition takes place in constant interaction with the state of the body into which it is embedded [23].

One important feature of embodiment, namely, that the interaction between body and cognition is subject to circular causality has been described by Wolfgang Tschacher [24]: The body acts as a control parameter on cognition and thus causes the generation of cognitive patterns. An example of this connection is the relation between facial expression and mood. Do I frown because a complaining voice is getting on my nerves or is the whining getting on my nerves because I am frowning? Maja Storch's examples of body feedback – that is to say, the feedback our psyche receives from our bodies – comprise connections from almost all forms of activity [25]. Do we sit bent over with drooping shoulders because we are about to give up on a difficult task or do we give up easily because we are sitting in a bent position? The feedback our body posture gives our psyche has been studied extensively. One experiment primed students by having them sitting for 8 minutes either in a crouched-up position, bent over forward with hanging head and shoulders, or in an upright one (spine erect and head lifted). After that period, both groups were asked to perform the impossible task of assembling a jigsaw puzzle from which pieces were missing. It turned out that the group which had been sitting crouched-up before the test gave up much more easily (they abandoned 16 puzzles on average) than the group which had sat in an upright position (which abandoned an average of just over 9) [26]. Another example of body-feedback is the connection between gait and mood: I am bouncing along the street because I am happy and I am happy because I am walking with a spring.

It is good news that we can influence our cognition through changing our body posture, our facial expressions or our breathing rhythm. To be aware of such co-variance is to be aware, too, of part of our usually transparent interfaces which influence our motorics and cognition. This awareness also enables us to recognize distortions in perception. In Chapter 7, we shall have a closer look at the way moods influence our judgement and how we can rid ourselves of unwanted influence by consciously separating perceptions and moods. So, possibly, Weibel's apprehension that it might turn out too difficult to reveal correlations between observer and observations can at least partly be dispelled if some of the usually hidden structure of our observer-participant perspectives are uncovered. The circular causality of body-psyche-feedback reveals strong co-variance, which can be described, conveyed and predicted.

Embodiment can be strong or weak, complex or less complex. Metzinger differentiates between First-, Second- and Third-order embodiment: First-order embodiment can be found in Artificial Intelligence which is based on bottom-up approaches (upward causation), for instance, in biorobots:

> The basic idea is to investigate how intelligent behaviour and other complex systems properties which were previously termed 'mental' can naturally evolve out of the dynamical, self-organizing interactions between the environment and a purely physical, reactive system that does not possess anything like a central processor (...) [27]

Second-order embodiment refers to a representational system, which "has a single, explicit and coherent self-representation of itself *as being an embodied agent.*" [28]. Examples are advanced robots, primitive animals, sleepwalking humans or individuals experiencing *le petit mal* (epileptic absence seizures). Third-order embodiment is inherent in a physical system which explicitly models itself as an embodied being and maps some of the representational content onto conscious experience – in other words, a system which consciously experiences itself as embodied and has a phenomenal self-model. Examples are humans in wakeful states or orang-utans moving from branch to branch.

For the central theme of this book – the structure of our fractal temporal perspectives – the transition between second- and third-order embodiment is most relevant because it correlates with the transition from invisible to visible underlying structures. (Metzinger has defined a similar notion, namely, the 'transparency constraint', which results from introspective unavailability and is inherent in phenomenological representations only – as implied in Metzinger's wolf example which we mentioned earlier). Of particular interest in this context is how usually invisible interfacial structures can be made visible by revealing covariance or other temporal distortions such as time dilation or contraction. Fractality comes into third-order embodiment through the fact that we are dealing with living beings which consciously experience themselves as embodied. Their cognitive processes are embedded in a physical body which is situated – that is to say, nested – in an environment. A nesting cascade can be continued by embedding the individual into a society, and the society into a larger level of organization such as mankind as a whole, or even the entire universe (we are stardust after all).

So far, we have been concerned with embodiment only insofar as cognition and the body itself are situated in a context. However, the fractal embedding structure also unfolds in the physical internal levels of the body. As Otto van Nieuwenhuijze suggests, the contextualization or nesting cascade extends in both directions: inwardly, from the level of the whole body to the levels of organs and cells, and outwardly, from the body level to that of the embedding social group [29]. We may think of humanity as being composed of humans just as our bodies are composed of cells. And just as human beings communicate, our cells communicate with each other, thus linking matter with in-formation. In this way, we create a body of knowledge on nested levels of description (LODs). Van Nieuwenhuijze points out that sensation of our context is always sensation of cells in our body. We – that is to say, our body cells – compare the impulses of the cells of the surface with sensations of cells in our body. So what we refer to as the world "around us" – our context – exists, in fact, as a thought image within us. This also means that all communication is internal communication on the respective LODs. Communication between individuals is also internal – situated within humanity, which can be seen as an organism in its own right. This means that all observations are also internal realizations – the universe as our mirror. Depending on our degree of involvement in our context, according to van Nieuwenhuijze, we have different levels of consciousness. And depending on our degree of identification with our body, other individuals, groups and our species, we have different degrees of awareness. And as awareness of context is based on the activity of our cells, we experience everything within our own bodies – as true endo-observers. The first casualty of this worldview is objectivity: The mode in which we observe depends on our mode of participation – that is to say, the degree of internal complexity we display and the number of simultaneous levels on which we interact with our context.

What results from this process is a cascade of nestings of modes of participation – a fractal structure. These nested nestings act as context for each other and cause simultaneous contrasts between the nested and the embedding structures and processes. If we look at temporal structures such as nested rhythms (or oscillations – I shall use these terms synonymously), the simultaneous contrasts manifest themselves either as phase-locking (entrainment) or more complex interference patterns. We shall see in the next paragraphs that successful phase-locking can be a sign of healthy adaptation but also that this is by no means always the case – it can also be a sign of a pathological condition.

It will be useful here to describe the difference between ordinary interference of rhythms (or oscillations) which do not necessarily lead to entrainment and interference which necessarily ends up in phase locking. For this purpose, we can cite the work of Strogatz, who set out to establish a "science of sync" – that is to say, the study of coupled oscillators. Examples of collections of oscillators include groups of fireflies, pacemaker cells and planets. Certain oscillators cycle spontaneously and repeat their rhythm over and over again at roughly the same intervals:

Fireflies flash; planets orbit; pacemaker cells fire. Two or more oscillators are said to be coupled if some physical or chemical process allows them to influence one another. Fireflies communicate with light. Planets tug on one another with gravity. Heart cells pass electrical currents back and forth. As these examples suggest, nature uses every available channel to allow its oscillators to talk to one another. And the result of those conversations is often synchrony, in which all the oscillators begin to move as one. [30]

The phenomenon of thousands of fireflies blinking on and off in unison is a well-known form of mass synchrony. It is thought to be initiated and maintained by an internal, re-settable oscillator, which has not yet been found, but it is thought to consist of a cluster of neurons in the firefly's brain. Each firefly in a large group continuously sends out and receives signals in the form of light flashes. Each emitted signal both shifts the rhythms of other fireflies and is, in turn, itself shifted by the flashes received. Synchronization occurs without an external conductor who co-ordinates the individual rhythms – the fireflies self-organize their rhythms through mutual cueing. If you have not yet witnessed this beautiful spectacle *in situ*, it is well worth watching a video clip on synchronized fireflies [31].

Strogatz points out that the oscillators can be either synchronized or desynchronized, depending on whether they are at the beginning, middle or end of a cycle when they are fired at [32]. In the case of pacemaker cells, for example, synchronization of coupled oscillators is desirable because it is the thousands of pacemaker cells, which synchronize their firings, that make our hearts beat. This is very similar to the fireflies: in each case, large populations of oscillators fire off pulses which interfere with the rhythms of the other oscillators in the group by either speeding them up or slowing them down. Both with fireflies and with pacemakers in the heart, synchronization is inevitable – that is to say, after a while, all oscillators phase-lock.

Fireflies and pacemaker cells always end up in sync when they function properly. This is vital in the case of cardiac pacemaker cells, since no heartbeat would come about if the cells fired at different times. Most oscillators, however, are rather more complex than insects or individual cells – as is found in the case of menstrual cycles. As Strogatz nicely put it: "Like women, most oscillators sync in some circumstances and not in others" [33]. Although some studies of how women's menstrual cycles shift in the presence of other women show a synchronizing effect, others detect no influence whatsoever or even antisynchrony [34]. One study shows that, if swabs with pheromones gathered from the sweat glands of a female donor are transferred onto another female's skin, this stimulates a shift in the recipient's cycle in a predictable way:

> Swabs taken from women at the beginning of their cycles, in the follicular phase before ovulation, tended to shorten the cycles of the women who received them. In other words, the recipients ovulated several days earlier than they would have otherwise, based on their prior records. In contrast, swabs taken from women at the time of ovulation prolonged the cycles of the beneficiaries. [35]

So, according to Strogatz' study, women who do not lead a hermit life are constantly shifting each others' cycles unconsciously – an interference which can but does not necessarily result in synchrony. We may say that, whether contact results in phase-locking or in more complex interference patterns, each woman is another woman's context. And if synchrony occurs, it is self-organized – that is to say, no external pacemaker is necessary apart from the women's mutual contextualization.

For both sexes, our circadian rhythm is an example of being at sync with our environment. In this case, the phase-locking of our sleep-wake cycles is a manifestation of our entrainment to the Earth's spin and the resulting daily cycle of light and darkness.

Being in-sync with one's immediate and/or remote embedding temporal rhythms can be of advantage. When male fireflies blink in unison and thus attract attention more effectively, it is conceivable that this will make mating more likely. On the other hand, unwanted phase-locking can lead to anything from a slight nuisance to a life-threatening condition. Inside the observer, entrained firing of neurons can lead to pathological feedback in the brain circuitry – in other words, to epileptic seizures. Although some epileptics respond to subcutaneously implanted pacemakers, which can suppress some seizures by sending electrical impulses through a wire connected to the vagus nerve in the neck, they are not as successful as the brain-pacemakers used to control Parkinson's disease [36].

A less dramatic but also potentially impairing common problem is audible feedback oscillation in hearing aids. This happens when the output of the receiver leaks into the microphone of the device and is re-amplified again and again, together with additional sounds the device picks up from the environment. Modern hearing aids have feedback cancellers which interrupt the unwanted feedback loop by, for instance, introducing an additional signal which cancels out the leaking sound [37].

Fig. 5.2. A simple version of a coupled oscillator.

As Alva Noë has pointed out, perception is action [38]: Whether or not the observer physically navigates through space, he interacts with the world and thus causes interference patterns. The same is true for temporal perception and cognition: Internal and external simultaneous rhythms cause temporal interference patterns which may or may not lead to entrainment. Most of these rhythms are nested, since the lengths of the simultaneous cycles usually vary.

Nested simultaneous rhythms require more than one temporal dimension for their description and are therefore temporal fractals. Succession (Δt_{length}) does not suffice to describe a nested system – embeddedness requires also simultaneity (Δt_{depth}). And as the embedding rhythms often extend both into the past and the future, both *causa efficiens* and *causa finalis* have to be taken into account when we deal with temporally nested systems. We mentioned the case of the missing fundamental as an example of such a temporal fractal in Chapter 3. In general, we may say that coupled and nested oscillators generate both Δt_{depth} (simultaneity) and Δt_{length} (succession).

When dealing with coupled or nested rhythms, we should also consider a qualitative distinction which goes back to antiquity: the notions of *chronos* and *kairos*.

Fig. 5.3. Saturn by Peter Paul Rubens, 1636. The painting depicts Chronos eating his own child (Saturn is the Roman name for the Greek god Chronos).

According to Greek mythology, the figure of *Chronos* arose from primordial chaos and has henceforth personified clock-time. Turning the Zodiac Wheel, he generates succession without assigning a special quality to any of the successive events.

In chronological time, events are arranged in a linear way, like beads on a string. *Kairos*, on the other hand, refers to the opportune moment in which the future may be influenced. As the god of the fleeting moment, *Kairos* personified a favourable opportunity. Whereas *chronos* is quantitative in essence, the *kairos* represents a bifurcation point which potentially gives rise to a new quality [39].

Fig. 5.4. Kairos: 16th century fresco by Francesco Salviati.

In the terminology of fractal time, *chronos* can be described as mere succession on one temporal LOD. The notion of *kairos*, however, requires an additional temporal dimension, namely, simultaneity in the form of nested LODs. Both simultaneous and nested rhythms give rise to the *kairos*. Self-organizing systems, such as the populations of fireflies which flash in unison, require two LODs: the level of the individual agents and the level of collective behaviour. The collective flashing is the emergent phenomenon – that is to say, the new quality of large-scale attention-arousing signalling, which adds up to more than the sum of the blinking of the individual fireflies in that population.

Simultaneous or nested temporal LODs are a necessary but not a sufficient condition for the *kairos*. The *kairos* also requires phase-locking. This phase-locking may occur on one level, when coupled oscillators like fireflies entrain to each others' rhythms. It may also occur between two or more LODs, as in the example of the overtones, where integer multiples of the fundamental frequency "vertically" locked into each other and formed a nested oscillator. Such nested simultaneous rhythms which phase-lock can be described in terms of Δt_{depth} and Δt_{length}. Δt_{length} refers to the number of individual oscillators on one LOD and Δt_{depth} to the number of embedding levels.

Nested LODs often exert both upward and a downward causation. Not only do shorter, embedded temporal structures influence or bring about longer, embedding ones, but also vice-versa: Longer rhythms influence shorter, embedded ones, such as the slow oscillations described by Buzsáki, which modulate faster local events (see Chapter 3).

The dynamics of both the observer and his context run together in the *kairos*. From an internal perspective, it becomes next to impossible to distinguish between a Now within which we can causally interact with embedding LODs and the phenomenon Jung called synchronicity: the temporally coincident occurrence of acausal events [40].

In nested systems, upward and downward causation can be interpreted as inside-outside causation – that is to say, the inner nestings (the shorter cycles) have a causal impact on the outer ones (the longer, embedding cycles) and vice-versa. (This is not true for strange loops, because circular causality is limited to a small number of LODs. The decisive upward causation which would disentangle the strange loop is blocked by the inviolate level.) Causality at work in the temporal dimension of Δt_{depth} describes a causal relationship which is referred to as a coincidence of acausal events if the LODs involved cannot be identified.

If, as van Nieuwenhuijze claims, all communication is internal (on the respective LOD), it is tempting to look at what happens when we choose a context which contains all frequencies at equal distribution – in other words, a context which can mirror all internal dialogue. This context is white noise, a random signal – that is to say, uncorrelated in time – which contains all frequencies at equal distribution. It is an ideal construction, of course, as real noise cannot have infinite bandwidth. However, one can produce white noise for a limited frequency band, which displays almost the same characteristics. White noise is a temporal fractal. It is statistically self-similar in the sense that its deviations remain the same under time dilation or contraction. So what does white noise do as context? When we stand close to a boiling kettle or the cooker-hood - devices whose bandwidth more or less resembles that of white noise – we may suddenly hear the phone ringing, the doorbell, the our mother's voice or any other familiar sound. The external signal which reaches us is extremely rich: ideal white noise contains all frequencies at an equal rate. So it is conceivable that we can trigger whatever response we expect.

This phenomenon was exploited in the Ganzfeld experiments – whose name derives from a German term meaning "entire field". The idea was to filter out all external signals, whether visual, auditory or haptic, in order to test individuals for extrasensory perception – an early version of isolation tanks. By filtering out environmental impacts, sensory deprivation was simulated and it was hoped that the test subjects would then hear internally generated impressions. Most results were inconclusive. However, the experiment is of interest in the context of our topic of embedded observer-participants because it produces the most fertile context imaginable: Anything can be extracted from or projected onto white noise.

There are colours of noise. By filtering white noise, we can produce pink noise, also known as $1/f$ noise – another temporal fractal, which manifests itself as correlations between fluctuations on nested time scales. It is present, for instance, in many musical tunes, heartbeat rhythms, meteorological, seismological and financial data. But although it seems to be ubiquitous in nature, the physical origin for pink noise is not known [41]. This makes pink noise a suspicious context: A pattern we observe everywhere may well be self-generated, possibly the result of a superimposed interfacial structure [42]. Depending on where we set the interfacial cut, pink noise can be assigned to the external system under observation, the measuring chain, the observer himself or any combination of these. One contributor is our brain: As Buzsáki has shown, the brain "generates a large family of oscillations whose spatio-temporal integration gives rise to 1/f statistics" – pink noise [43]. The neuronal $1/f^\alpha$ noise hints at both most complex dynamics but also high instability. As Bak et al have pointed out [44], 1/f systems self-organize without external tuning and undergo phase transitions – a behaviour they coined "self-organized criticality" (SOC). However, in the brain, SOC (which, for instance, in the shape of neuronal avalanches, may cause epileptic seizure) is usually prevented by stabilizing oscillatory dynamics. Also phase-locking between neighbouring cortical oscillators is prevented by the fact that the ratios between their mean frequencies are not integers, so adjacent bands cannot linearly phase-lock [45].

As we saw earlier in this chapter, phase-locking of some oscillators within our bodies (e.g. heart cells) is vital, whereas others would be disastrous (e.g. supersynchronous activity of neurons which manifest themselves as epileptic seizures). This is also true for "inside-outside phase-locking" – that is to say, entrainment between observer and context. Apart from entrainment into rhythms such as the spin of the Earth, we also entrain to other individuals' mental states by simulating both the physical side we observe and the corresponding mental state (e.g. a crouched posture and the willingness to give up easily on a task, or a bouncing gait and a happy mood). This simulation is carried out by mirror neurons – nerve cells which fire when we simulate another individual's behaviour and mental states. Mirror neurons were detected by Giacomo Rizzolatti in the early 1990s [46] and are now studied also in the context of education, since learning by imitation can now be explained in a new light.

Recent research has shown that we are capable of resisting such "contextualization by mirroring". Cheng et al's results show that individuals can resist contextualization. for instance, in the case of doctors unlearning to empathize with their patients' plights [47]. Self-protecting mechanisms of this kind would allow such a doctor to act out a role without inflicting damage on himself. However, if taken too far, such desensitizing can have horrific consequences, for instance, when it produces individuals who are capable of torturing others. The reality which we generate and discern is the reality of contextualization. Therefore, an ethics of constructivism must place the right to resist contextualization at the centre of discussion.

Contextualization can be measured in terms of our degree of involvement. We can widen our perspective by contextualizing or narrow it by resisting contextualization. If our ability to contextualize is compromised, we cannot generate simultaneity – like the non-fractal observer described in Chapter 2.

The Now is the culmination point of successful and unsuccessful contextualization – a temporally extended window, in which time and duration emerge as the result of simultaneous contrasts. Simultaneity is generated by a successful handshake – meaning that an interaction takes place. Succession is generated by an unsuccessful attempt to interact. However, success here should not simply be equated with positivity. (Given our terminology, epileptic seizures would be classed as internal successive handshakes.)

I have addressed both cognitive and physical intromission and extramission in a phenomenological framework, exemplified by nested oscillators, mindfulness and embodiment. The notion of embodiment, in particular, requires a communication model which takes account of both cognitive and physical intromission and extramission. A promising candidate is Jerry Chandler's model, which includes novel concepts such as exformation [48]. In a nutshell, the idea is that real flesh-and-blood human beings exchange emissions which undergo a number of transformations on the way from person A to person B because multiple message sources have to be integrated. What we perceive are sensory impressions, which first have a physical impact on us, say, air pressure waves we perceive via our auditory perceptual apparatus and our epidermis. Next, these impacts are transformed into information by our bodies. We then abstract from this information what is meaningful to us in terms of the symbol systems at our disposal. Our reaction to these incoming impressions and the transformations they undergo is that we conclude that we have understood something and wish to respond. In order to do so, we infer from the outcome of our internal translations and re-start the procedure – this time in reverse. We extract what we regard as relevant from available symbol systems and formulate a response which is expressed via our sensory-motor system as sound waves we produce and/or gestures and mimics. Given the many transformations along the line, a lot of the communicative process will remain opaque to the parties involved.

Another interesting feature of the interaction of intromission and extramission is the extended boundary resulting from the interference pattern of inside and outside, which manifests itself as 'potential space' – that is, if we include cognitive extramission in the shape of intentions and goal-directed behaviour. Developmental psychologist D.W. Winnicott defines this extension – the potential space – as a realm which belongs neither to the observer nor to the rest of the world – it is a sphere of influence which partakes in both [49]. Potential space is both created and inhabited by what Winnicott refers to as *transitional objects*. He describes these as the first non-I: for instance, a teddy bear, a security blanket like that carried around by the cartoon character Linus, or a lullaby which an infant may hum. Their function is to familiarize oneself with the unknown by mediating between inside and outside, self and non-self.

Transitional objects (the objects used) and phenomena (the techniques employed) function as a special kind of context. They do not belong only or the inside or the outside, but partake of both, not unlike internal and external interacting rhythms. Together, they form interfaces – that is to say, interactive filters with the world, and thus govern how we link inside and outside. The potential space both joins and separates me, me-extensions, and the non-me. Play and cultural experience, as well as science and religion, belong to this realm between subject and object, between inside and outside.

As a developmental psychologist, Winnicott is concerned with the acquisition of transitional objects and potential spaces in early infancy. The mother adapting to the infant's needs plants and reinforces the illusion of magical control over the world. It is only when the child gradually realizes that it is not in control that its first concept of an external reality emerges. And it is the transitional object which forms the first interface with the rest of the world. Although people around the child will recognize that object as being a teddy bear or a blanket, they are aware at the same time that it is far more than this. The meaning that the infant attributes to its first possession makes it the first object it has created: a link to external reality. Adults create potential spaces too. Their transitional objects can be found in culture, art, religion and science. But also objects like teddies or dolls are often re-adopted late in life to compensate for a loss or an unfulfilled need. Some transitional objects stay with us for many years, some for a lifetime. More often, however, they tend to be of an ephemeral nature and become obsolete, ending up as mere favourite objects. Their original meaning and function have been incorporated in the owner once the object is no longer needed – through a boundary shift towards the inside, it has become assimilated context.

So we may contextualize by 'weaving together' inside and outside within potential space. The resulting interference pattern which forms our boundary, our interface reality, has an extended structure, which hosts transitional objects and phenomena.

A similar boundary shift occurs when we experience a change in our proprioception. Glaser has given the term *transsensus* to the ability to extend our self-perception to other living beings or objects by integrating them into our own proprioception [50].

Fig. 5.5. *Transsensus*: Merging boundaries of horse and rider. [51]

The process of incorporating others or melting into an object forms transcending boundaries. When we ski down a piste, for instance, we "feel" the consistency of the snow as being powder-like or icy through the way our skis sink into the snow or the amount of friction created. Our proprioception has changes in such a way that skier and skis have become one systemic whole, through which the world is perceived [52]. Transsensus also emerges when we imagine that parts or all of another person's (or animal's) body belongs to our own. For example, if a physician's hand is resting on our skin, and we think of that hand as being part of ourselves, then our physical state can measurably improve: better circulation, deeper breathing and an increase in the elasticity of our tissues. Closely connected to this skill of transsensus is the ability to trust. Kestenberg describes a "trusting" baby-belly as an example of this attitude of trust which triggers deep and smooth respiration [53].

Each person is another person's context. But it is largely up to us whether we allow parts of ourselves to be assimilated, as was the case in the borderline patient in Chapter 4, or decide to integrate parts of our environment or other human beings into our proprioception. Transsensus can be hell if it forces us into a rigid context (*L'enfer c'est les autres* [54]) or limited by inflexible frameworks (trapped by category). On the other hand, it can be heaven if the incorporation of another person into our proprioception allows us to experience love and beauty. *Homo homini contextus*.

References

1. Etching depicting visual rays emanating from the eye in the shape of a cone (unknown Renaissance artist).
2. I. Putri, *Ancient Theories of Vision and Al-Kindi's Critique of Euclid's Theory of Vision*. Proseminar: History of Computational Science, Vision and Medical Science at the University of Applied Sciences, Munich, Germany, Lehrstuhl für Informatikanwendungen in der Medizin & Augmented Reality 2007.
3. M. Velmans, Are we out of our Minds? in: *Journal of Consciousness Studies*, Vol. 12, No. 6, *Special Issue: Sheldrake and his critics – The Sense of Being Glared at*. June 2005.
4. O.E. Rössler, How Chaotic is the universe? in: *Chaos*, Ed. by A.V. Holden, Manchester University Press 1986, pp. 315-321.

5. G.W. Leibniz, *Monadologie.* Ed. by Hartmut Hecht, Stuttgart, Reclam 1998.

6. B. Gick and D. Derrick, Aero-tactile integration in speech perception, in: *Nature*, Vol. 462, Nov. 2009, pp. 502-504.

7. P. Weibel, The Noise of the Observer, in: *Ars Electronica: Facing the Future.* Ed. by Timothy Druckerey, MIT Press, Cambridge 1999, p. 6.

8. P. Weibel, 1999, p. 7.

9. D.R. Hofstadter, *Gödel, Escher, Bach - An Eternal Golden Braid.* Vintage Books 1980.

10. O.E. Rössler, personal communication 2007.

11. P. Weibel, 1999, p. 7.

12. T. Metzinger, *Being No One: Consciousness, the Phenomenal Self and the First-Person Perspective.* Foerster Lectures on the Immortality of the Soul. The UC Berkeley Graduate Council, 2006.
http://video.google.com/videoplay?docid=-3658963188758918426&q=neuroscience and personal correspondence 2009.

13. A. Moore, Transactional Fluctuations I. In: *Variantology 3; On Deep Time Relations of Arts, Sciences and Technologies*; Ed. by Siegfried Zielinski & Eckhard Fürlus. Verlag der Buchhandlung Walther König, Cologne 2007 and personal communication 2009.

14. D.J. Simons and D.T. Levin, Failure to detect changes to people during a real-world interaction, in: *Psychonomic Bulletin & Review*, 5 (4), 1998, pp. 644-649.

15. D.J. Simons and C.F. Chabris, Gorillas in our midst: sustained inattentional blindness for dynamic events, in: *Perception*, Vol. 28, 1999, pp. 1059-1074.

16. E.J. Langer, *Mindfulness.* Da Capo Press, Cambridge, 1989.

17. E.J. Langer, 1989, p. 10.

18. Noise-cancelling technology: Opting for the quiet life, in *The Economist*, February 13th-19th 2010, pp. 76-77.

19. B. West, *Where Medicine Went Wrong. Rediscovering the Path to Complexity.* World Scientific, Singapore 2006.

20. P. Davies, The Physics of Downward Causation, in: *The Re-Emergence of Emergence*, Ed. by P. Clayton and P.S. Davies, Oxford University Press 2006.

21. P. Davies, 2006, p.6.

22. D.R. Hofstadter, 1980, p. 709.

23. W. Tschacher, Wie Embodiment zum Thema wurde, in: Maja Storch et al: *Embodiment Die Wechselwirkung von Körper und Psyche verstehen und nutzen.* Huber, Bern 2006, p. 31.

24. W. Tschacher, 2006,

25. M. Storch, Wie Embodiment in der Psychologie erforscht wurde, in: Maja Storch et al: *Embodiment Die Wechselwirkung von Körper und Psyche verstehen und nutzen.* Huber, Bern 2006, p. 31.

26. J.H. Riskind and C.C. Gotay, Physical Posture: Could it Have Regulatory or Feedback Efects on Motivation and Emotion? in: *Motivation and Emotion*, Vol. 6, No. 3, Sept 1982, Springer Netherlands pp. 273-298.

27. T. Metzinger, First-order embodiment, second-order embodiment and third-order embodiment, in: Self-models, *Scholarpedia*, www.scholarpedia.org/article/Self_models, 29.12.2009.

28. T. Metzinger, 2009.

29. O. van Nieuwenhuijze, personal communication 13 February 2010.

30. S. Strogatz, *Sync – How order emerges from chaos in the universe, nature and daily life*. Hyperion, New York 2003, p. 3.

31. Synchronized fireflies: www.youtube.com/watch?v=a-Vy7NZTGos

32. S. Strogatz, 2003.

33. S. Strogatz, 2003, p. 38.

34. M.K. McClintock, Menstrual synchrony and suppression, *Nature* 229, 1971, pp. 244-245.

35. S. Strogatz, 2003, p. 37.

36. B. Johnson, Pacemaker may avert epileptic seizures, *The Guardian*, 27 June 2006.

37. I. Merks, S. Banerjee and T. Trine, Assessing the Effectiveness of Feedback Cancellers in Hearing Aids, in: *Hearing Review*, April 2006.

38. A. Noë, *Action in Perception*, MIT Press, Cambridge 2004.

39. S. Vrobel, The Kairos Syndrome, in: *International Journal of Computing Anticipatory Systems*, Vol. 20, Ed. by D.M. Dubois, CHAOS, Liège, Belgium 2007, pp. 153-162.

40. C.G. Jung, *Über die Psychologie des Unbewußten* (1943), Fischer, Frankfurt a. Main, Germany 1981.

41. R.F. Voss, Fractals in Nature: From Characterization to Simulation, in: H.O. Peitgen and D. Saupe (Eds.), *The Science of Fractal Images*. New York, Springer 1998.

42. S. Vrobel, How to Make Nature Blush: On the Construction of a Fractal Temporal Interface, in: *Stochastics and Chaotic Dynamics in the Lakes: STOCHAOS*. Ed. by D.S. Broomhead, E.A. Luchinskaya, P.V.E. McClintock and Tom Mullin. American Institute of Physics, AIP Conference Proceedings 502, 2000, pp. 557-561.

43. G. Buzsáki, *Rhythms of the Brain*. Oxford University Press, Oxford, U.K., 2006, p. 130.

44. P. Bak, C. Tang and K. Wiesenfeld, Self-Organized Criticality, *Phys. Rev. A*, Vol. 38, Issue 1, 1987, pp. 364-374. 1987.

45. G. Buzsáki, 2006, pp. 134-135.

46. G. Rizzolatti, *Mirrors in the Brain – How Our Minds Share Actions and Emotions*. Oxford University Press 2008.

47. Y. Cheng, C. Lin, H. Liu, Y. Hsu., D. Hung, J. Decety, Expertise Modulates the Perception of Pain in Others. In: *Current Biology*. Sept. 2007, pp. 1708-1713.

48. J.L.R. Chandler, *Biosynthesis of Consciousness in Poised Perplex Systems*, Talk held at CASYS '09 in Liège, Belgium August 2009.

49. D.W. Winnicott, *Human Nature*. Free Associations Books, London 1988, pp. 101-107.

50. V. Glaser, *Eutonie. Das Verhaltensmuster des menschlichen Wohlbefindens*. Lehr- und Übungsbuch für Psychotonik, 4th Edition, Karl F. Haug Fachbuchverlag, Heidelberg 1993.

51. Photograph courtesy of Mulemwa Mususa 2009.

52. G. Weber, *Kraniosakrale Therapie*. Springer, Vienna 2003, p. 54.

53. J.S. Kestenberg, Transsensus-Outgoingness and Winnicott's Intermediate Zone, in: *Between Reality and Fantasy. Transitional Objects and Phenomena*, Ed. Simon A. Grolnick and Leonard Barkin, Jason Aronson Inc., pp. 10 ff.

54. J.-P. Sartre, *Huis Clos*. Gallimard, Paris 1947.

Chapter 6

Temporal Binding:
Synchronizing Perceptions

The smell of an over-ripe papaya at my local supermarket recently brought back to me the memory of a trip through the Central American jungle. Immediately, the entire activation pattern was firing again in my brain: I re-heard the sounds of the howler-monkeys, felt the sticky surface of moist plants on my skin, the high humidity which had taken away my breath. Smell is a most evocative trigger, which can instantly carry us back even into early childhood memories. But such memories can also be triggered by any one or combination of the other modalities which were present during their formation. How come? All the sensory impressions which reach our brain have a specific arousal pattern, which is relayed to the associative cortex. There, it activates already existing nerve cell couplings, which have been formed by earlier sensory impressions. If these impressions vary slightly, the superposition of both the old and the new arousal pattern will form a new, extended activation pattern which corresponds with the newly integrated experience [1]. The more often the activation pattern is activated, the stronger it becomes: Cells that fire together wire together. And if – as is usually the case – the activation pattern has been formed by a combination of different modalities such as visual, auditory, olfactory and tactile stimuli, the entire pattern may be re-activated and fire as a result of a single stimulus arriving in the brain. The more which modalities were involved when the pattern was formed, the more stable the memory, since it can be re-activated via several channels.

This insight into the way we turn simultaneous multi-modal stimuli into a coherent perceptional gestalt by stabilizing them in the shape of coupled cells is a fairly recent one. It evolved from the early conditioning

117

experiments of Ivan Pavlov [2]. However, the conclusion we can now draw – namely, that memories are laid down in a more stable way if a number of perceptual modalities are involved – was already known to Johann Heinrich Pestalozzi, a German educationist, who proclaimed that children should "learn with all their senses" [3]. Not only do Pestalozzi schools all over the world base their curriculum on this principle, but also other anthropological and holistic learning institutions.

In the last chapter, we said that embodiment works because body posture is coupled with emotions, moods or attitudes. Synchronous arousal patterns, which manifest such entire perceptional gestalts, composed of motoric, sensory, affective and cognitive parts, are known as "distributed assemblies" [4]. If we are in the habit of meeting our friends every Saturday night at a beer garden, where we eat, drink, chat among linden trees, dance and listen to the local band, all these sensory inputs turn into a perceptional gestalt which, on the neural level, manifests itself as a synchronous arousal pattern. After a while, a single trigger, say, a familiar piece of music or the scent of the linden trees will suffice to stimulate the desire for a drink. This conditioning effect, which is caused by distributed assemblies firing synchronously, is the reason why addiction is so hard to overcome.

However, there are also positive ways in which such trained collective firing can be applied, as when children are encouraged to learn with all senses. Not only does this enrich learning and make it more comprehensive, but it also facilitates memory retrieval, as there are several pathways which lead to the same arousal pattern.

But how do we mould all these different sensory inputs into one meaningful experience in the first place? The *binding problem* of perception addresses the question of how we are capable of combining, within a window of approximately 100-200 milliseconds, such various features as colour, texture, smell and distance into the representation of an object [5]. Although different neurons in separate parts of the cortex are involved in processing visual, auditory or tactile stimuli, which arrive at different speeds, the brain somehow manages to generate a coherent representation from these cross-modal stimuli. Until fairly recently, no convincing explanation could be given for what actually happens on the neural level when we form a gestalt – that is, a coherent representation of an object from a number of stimuli. One possible explanation, proposed in the 1980s, is to be found in coupled gamma oscillations. Gamma waves have a frequency of typically 40 Hz and are ubiquitous in the human brain. When they phase-lock, they trigger temporally synchronous discharge in separate parts of the cortex [6]. This model, known as *binding by synchrony*, is based on simultaneity – that is to say, on temporal congruence rather than spatial connections. In this model

Convergence of connectivity is no longer the main variable of feature extraction; rather, it is temporal synchrony of neurons, representing the various attributes of objects, that matters. The different stimulus features, embedded in the activity of distributed cell assemblies, can be brought together transiently by the temporal coherence of the activated neurons. [7]

German neuroscientist Wolf Singer suggests that this synchronization of neuronal oscillations is a necessary condition for consciousness:

One candidate mechanism for dynamic binding is the precise synchronization of neuronal responses that occurs when neuronal populations engage in well synchronized oscillatory activity in the beta- and gamma-frequency range. (...) These synchronized oscillations are strongly reduced or missing when the brain is in states that are incompatible with conscious processing. [8]

As we have seen earlier, phase-locking occurs in the external world, between inside and outside (when human internal rhythms entrain to external ones) and within the body (for instance, in the brain during an epileptic seizure). So how can we tell whether gamma synchronization is "home-made" or imported from the external world? Singer describes an experiment in which synchronization in the visual cortex of cats correlated with perceptual grouping – that is to say the recognition of a gestalt. Synchronization was found within and among different areas of the brain, as well as between hemispheres, and proved to be genuinely "home-made":

> (...) none of these synchronization phenomena were detectable by correlating successively recorded responses to the same stimuli. This indicates that synchronization was not due to stimulus locking of responses but to internal dynamic coordination of spiking times. [9]

An interesting observation in this context is the fact that stimulus-related synchronization is particularly strong when the global electroencephalography desynchronizes – that is to say, when there is more noise or disorder in the brain's dynamics and when the individual, whether a person or an animal, is attentive. In fact, synchronization takes place even during the preparatory state of anticipating a task, which precedes the appearance of the stimulus and arouses attention. When cats were preparing themselves for a task and focussed their attention, several cortical areas synchronized with zero phase-lag. This synchronization increased immediately when the visual stimulus appeared and remained coordinated until the task was completed. After completion, synchronization disintegrated and was replaced by low-frequency oscillations which did not phase-lock. [10]

Singer correlates this anticipatory gamma activity with the reactivation of memory. He reports an experiment in which individuals had to recognize words, some of which they had registered and semanticised consciously and unconsciously. Just prior to and during the recognition of the consciously perceived stimuli, both theta and gamma activity occured. When unconsciously perceived stimuli were recognized, no co-ordinated oscillations were observed. This suggests that it is phase locking of oscillations which distinguishes conscious from unconscious processing in the brain:

> About 180 ms after presentation of stimuli that were consciously perceived, there was an epoch, lasting around 100 ms, during which induced gamma oscillations recorded from a large number

of regions exhibited precise *phase locking* both within and across hemispheres. [11]

Singer concludes that consciousness is not associated with activity in a particular region of the brain but is an emergent phenomenon triggered by precise temporal coherence among distributed neural assemblies. Short windows of perfect synchrony could act as pacesetters which provide a common temporal reference frame, into which events could be integrated. A promising candidate for such a time setter is the global theta oscillation which sets in after events that trigger conscious processing. This would mean that, together with gamma- and beta-oscillations, a fractal clock ticks away in the brain:

> (...) slow oscillations in the theta range have been found to be coupled by the coexisting beta- and gamma-oscillations. This suggests the hypothesis, that local coordination of computations within specific cortical areas is achieved by fast ticking clocks, such as beta- and gamma oscillations, while global and sustained integration of local results is achieved at a slower pace by low frequency oscillations. [12]

The more global the representation, the longer is the time scale for integrating distributed information. It is perhaps more than a mere coincidence that the duration of subjective presence corresponds approximately to the cycle time of theta rhythms. [13] (Theta rhythms oscillate at 6-10 Hz and occur in humans during drowsy and meditative states).

Singer also studied the connection between meditative states and the occurrence of synchronized gamma oscillations. In an experiment, he compared the synchronous activity in the brains of Tibetan monks, who had been in the habit of meditating for decades, with those of students who had never meditated before and found a surprising level of activity in the left frontal cortex of the monks when they went into deep meditation: On the neural level, meditation correlated with synchronously communicating gamma frequencies. Singer calls the gamma frequency (approximately 40 Hz) the "frequency of silence", because he believes that the synchronicity of gamma oscillations is triggered by the silence sought in meditation, when one attempts to rid oneself of all thoughts. [14]

The peaceful state of mind experienced in meditation, as well as the mental and physical feeling of being at one with oneself and with the world, can play a role in the treatment of depression and pain. It does so in a large number of medical institutions in the United States, which seek to influence the mood of patients so as to train them how to live with chronic pain. This can be done by focussing on breathing or a mantra.

The surprising extent of synchronous gamma oscillations which were observed by Singer during meditation is an example of internal entrainment, as opposed to entrainment between the human body and the environment. What happens during meditation is not nesting: although simultaneity is created, it is an exact locking-in of oscillations, a congruent phase of neurons firing at the same time. Against the background of Singer's findings, it is necessary to distinguish between two kinds of simultaneity, namely, nested and non-nested ones. Nested simultaneous events create a temporal perspective which spans several LODs, which generates depth. Entrainment, by contrast, creates a temporal perspective which is reduced to a few LODs, possibly only one. Focussing on breathing or a mantra and eliminating all thoughts and successfully ignoring external stimuli – all these narrow our temporal perspective. To a degree, this is what happened to the law enforcement officer described in Chapter 1, who experienced tunnel vision and reduced auditory perception.

Mindfulness is another way of reaching this temporal perspective of reduced LODs. Instead of letting everyday actions run automatically – that is to say, allowing neural assemblies to fire together – focussing on individual aspects helps to resist hard-wired responses and allows habitual patterns to be disentangled, both on the semantic and the neural level. However, this takes training and persistence. American neuroscientist Richard Davidson has shown that monks who had been in the habit of meditating for decades produced higher synchronicity in the brain than students who had been meditating only for a few months [15]. So this temporal perspective of diminishing LODs can be trained. John

Kabat-Zinn has developed a method which is known as *mindfulness-based stress reduction*. It is mainly used to treat chronic pain, stress and the disorders accompanying cancer treatment [16].

These findings support my suggestion that fractal and non-fractal observers perceive the world very differently. Fractal observers do so via a nested temporal interface, whereas a non-fractal observer's event horizon is reduced to one LOD. Generally, displaying a high degree of fractality – that is to say, a high level of internal complexity – is a sign of good health [17]. However, the skill of reducing the degree of temporal fractality is also of advantage in dangerous situations, when we experience stress, and in meditation, when, ideally, we focus only on one phenomenon and resist all contextualization. For the police officer, a narrow focus on essentials was vital: By blotting out all unnecessary perceptions, he was able to devote his full attention to the shoot-out. By humming a mantra or concentrating on breathing, an individual engaged in meditation will, ideally, enter a state in which it is possible to resist all contextualization and avoid disturbing influences.

Synchronization is one way of explaining the binding problem. However, the question remains as to what happens before synchronization: How do we merge into one meaningful event those cross-modal stimuli which take different amounts of time to reach our body and our brains? The simultaneity created by entrainment is not the same as that generated by temporally nested events. Entrainment leads to a reduction of LODs because all oscillations are congruent in time – in other words, although the oscillations may occur in different neural assemblies, they are *temporally* on one LOD. Nested events, on the other hand, cannot be reduced to one LOD because they overlap (unless they are perfectly self-similar – see Chapter 10). In terms of fractal time, we may say that, during entrainment, when we observe total congruence, Δt_{depth} approaches zero, whereas the processing of nested events retains the number of LODs of the overlapping stimuli: Δt_{depth} remains stable. Nevertheless, all these LODs are moulded into one event. How come? David Eagleman suggests that we coordinate the stimuli which arrive via our various senses – vision, hearing, touch, smell, etc. – by creating a time window within which we mould them into one event. Our brain makes up for speed disparities both between and within sensory channels: Within the visual system, for instance, by integrating colour, motion and spatial perspective into one object. He believes that our brain manages this integrative task by waiting for the slowest information to arrive. This time window happens to extend to about a tenth of a second:

> In the early days of television broadcasting, engineers worried about the problem of keeping audio and video signals synchronized. Then they accidentally discovered that they had around a hundred milliseconds of slop: As long as the signals arrived within this window, viewers' brains would automatically resynchronize the signals; outside the tenth-of-a-second window, it suddenly looked like a badly dubbed movie. [18]

Apparently, our visual system makes up for the delays. This means, of course, that our perception of the event does not occur in real time – we perceive an event which happened in the past. Eagleman thus concludes that awareness is postdictive. We integrate visual stimuli within a window of a tenth of a second and, in retrospect, make sense of them. For other senses, such as hearing or touch, the window may have a slightly different extension, but the underlying moulding mechanism is the same. Even within one modality, delay times are readily compensated by our brain:

> If you touch your toe and your nose at the same time, you will feel those touches as simultaneous. This is surprising, because the signal from your nose reaches your brain well before the signal from your toe. [19]

It seems that, within a time window, our brain combines information into one unified perception of the world, which is formed in retrospect. As Eagleman points out, "given conduction times along the limbs, this leads to the bizarre but testable suggestion that tall people may live further in the past than short people." [20] He compares perception with the airing of a live television show: We do not make sense of the world in real time but in retrospect, after a short delay, which allows to us to perform corrections if editing should be necessary to generate and maintain a non-contradictory world.

This time delay which results from our brain waiting for the slowest modality is another factor to be considered in the binding problem. To this can be added the fact that our brain collects and merges perceptions from different senses at different speeds.

But how come our brain seems to know what happens simultaneously in the external world? Eagleman provides a convincing answer: Our brain constantly recalibrates our expectations about arrival times. And the best way of anticipating the relative timing of incoming stimuli is to interact with the world. Whenever we touch or bump into something, our brains assume that all visual, auditory and tactile inputs we perceive within the small window of time are simultaneous. Thus Eagleman's hypothesis sheds new light on Noë's conclusion that perception is action [21].

We saw in Chapter 4 that being used to delay times makes us highly vulnerable to distortions of the temporal order when, for instance, a delay we have grown accustomed to is suddenly removed. Taking away the expected delay fools our brains into an illusory reversal of time, as happened in the example of the pressed key which is followed, after a short delay, by a flash. Removing the delay makes us see the flash before the key press. Anticipation is a powerful corrective and vital to our assignment of causality. Did I first hit the key or did the light flash before I pressed the button? To maintain a calibration which generates a non-contradictory world for us, it is necessary to continuously interact with our environment (which includes other people). This interaction is a reality check which normally prevents us from feeling disoriented. However, as Eagleman stresses, in certain disorders, interaction does not guarantee successful navigation:

We have recently discovered that a deficit in temporal order judgements may underlie some of the hallmark symptoms of schizophrenia, such as misattributions of credit ("My hand moved, but I didn't move it") and auditory hallucinations, which may be an order reversal of the generation and hearing of normal internal monolog. [22]

Similar observations have been made by Dieter de Grave, who claims that schizophrenia is a biopsychosocial order in its own right" [23], as all nesting of events follows a closed logic. It is only when a schizophrenic's expectations do not match those of other observer-participants, that contradictions arise, which may lead to conflicts. Laurent Nottale's and Pierre Timar's research also tackles this problem in terms of the congruence between inside and outside simultaneity [24]. When we look at certain forms of schizophrenia, we are really dealing with timing (or nesting) diseases. Unsuccessful contextualization leads to disorientation. We will deal with this topic in more detail in Chapter 9, when we look at *temporal misfits*.

If the brain waits for the slowest signal to arrive, as Eagleman suggests, all awareness is postdictive. The delay is extremely useful, as it allows for correctives, so that incoming stimuli can be adjusted, in order to meet our expectations and generate a non-contradictory image of the world.

We generate simultaneity within that tenth-of-a-second window when we perceive the world by interacting with it: perception is action. Whatever we then rate as simultaneous or successive determines the causal order of events of our realities and, thus, the internal logic of our worlds.

Your brain, after all, is encased in darkness and silence in the vault of the skull. Its only contact with the outside is via electrical signals exiting and entering along the super-highways of nerve bundles. Because different types of sensory information (hearing, seeing, touch and so on) are processed at different speeds by different neural architectures, your brain faces an enormous challenge: what is the best story that can be constructed about the outside world? [25]

The brain is continuously integrating stimuli in such a way that no contradictions arise. Depending on the way they are conditioned by experience, individual A will create very different realities to individual B. These realities will be logically coherent, though they may turn out to be mutually incompatible. The integrative process which is at work here involves the nesting of stimuli, in the wake of which a temporal perspective is generated. The creation of simultaneity through such nesting increases Δt_{depth} and decreases Δt_{length}. The opposite is true for purely internally generated simultaneity which manifests itself as entrainment in the brain. The synchronization of gamma waves, which have been associated with transcendental meditative states, are an example of such entrainment. During synchronization, stimuli are de-nested, which means that Δt_{depth} decreases and Δt_{length} approaches infinity. The same is true for entrainment between inside and outside, when the body locks into external rhythms [26].

When internal and external structures resemble each other, there seems to be a pleasing congruence, which is often explained by the observation that humans are biophilic – that is to say, they like other living creatures and natural shapes. There appears to be a further, strange attraction for humans in the shape of fractals, an appeal which we may call *fractalphilia*: the observation that humans find fractal structures appealing. Let us have a look at two very different pathways, both of which lead to *fractalphilia*.

The first pathway relates to neural synchrony as the basis of binding, as discussed earlier in this chapter. An experiment conducted by Erimaki et al shows that fractals are more efficient in evoking neural synchrony and binding than other stimuli such as Kaniza figures, which are commonly used to study perceptual binding (see Fig. 6.1) [27]. The EEG responses to a complex fractal stimulus such as the Mandelbrot set (see Fig. 6.2) showed an increased synchronization of the EEG activities over the parieto-occipital areas, compared to the degree of synchronization a Kaniza figure would evoke.

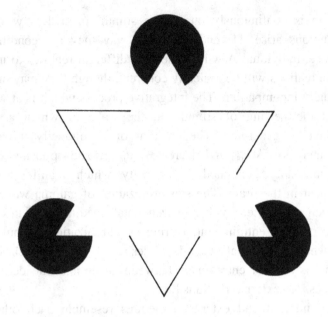

Fig. 6.1. There are many variations of Kaniza figures. They all have in common that the observer sees imaginary contours (the white triangle in the middle).

To compare the various degrees of synchrony, Synchronization Likelihood (SL), a measure of both linear and non-linear interdependencies between EEG channels, was employed. Erimaki et al observed an increase in SL not only for gamma-1 and gamma-2 bands, but also for broadband SL, which means the interdependencies were not tied to specific frequencies. Earlier in this chapter, we pointed out the correlations between consciousness and synchrony, as well as the meditating monks' feeling of total immersion and peacefulness. Although the increased synchrony measured during both the perception of appealing fractals and the peaceful state of meditation is a mere correlation, it is an interesting observation. Erimaki et al suggest that the increased synchronization they observed when individuals were exposed to fractal images may provide a neurophysiological basis for the frequent hypothesis that aesthetic preference is given to fractal images as opposed to non-fractal ones like the Kaniza figures.

Fig. 6.2. Zoom into the Mandelbrot set. [28]

Erimaki et al's results support the idea that, if the observer's fractal temporal perspective matches that of multi-sensory and multi-layered incoming stimuli, the internal degree of complexity roughly equals the external one. This results in a positive aesthetic perception. If both the internal and external structures display a similar degree of nesting, they may be congruent to a certain degree and synthesize a version of the world that is less complex. However, synchronization of nestings on the inside and the outside produces a kind of synchrony different from that which we have discussed in this chapter so far.

Which brings us to the second pathway to synchronicity – one that is concerned with matching internal and external patterns. Yannick Joye describes the advantages of implementing fractal geometry in architectural design, ranging from emotional to health effects, such as stress reduction [29]. As we mentioned earlier, the human body displays many fractal structures, from the scaling branching of our bronchial tubes to the fractal rhythms of our breathing, and so does our environment [30]. We appear to show a preference for low to intermediate fractal dimensions around 1.3, which matches the dimensions of natural objects such as clouds or coastlines. Our mind seems to be attuned to processing patterns with fractal dimensions of this scale (which roughly matches the level of complexity of the Koch-curve, a mathematically generated fractal with a fractal dimension of 1.26). This dimension is consistent with studies of the relationship between complexity and aesthetic preference. The mind appears to look for an optimal arousal level between highly complex and simplistic patterns [31].

But how come we prefer fractals of dimension 1.3 to more complex and more simplistic structures? One answer may be that such a fractal environment provides restoration from stress. Taylor has shown that an environment which provides a maximum of stress reduction also happens to display a mid-range fractal dimension around 1.3. This was revealed in a study which determined the relation between the level of fractality of a visual stimulus and the degree of stress reduction [32] Stress reduction was measured by skin conductance while participants looked at four images: a photograph of a forest, a simplified representation of a savannah (a landscape painting), a picture with squares and a control which consisted of a blank white plane. Taylor found that the painting of the savannah, whose fractal dimension matched the value of maximum aesthetic preference, namely 1.3, was most effective in reducing stress. This revealed that it was not the degree of naturalness of the depicted contents which determined the potential of an environment to reduce stress, but the fractal dimension. Otherwise, the photograph of the forest, which was far more realistic and natural-looking, would have produced the greater restorative response.

It is one thing to determine the fractal dimension which provides maximum restoration. A more difficult task is to explain why this is so. If this preference is a selection effect, we would need to propose the advantages of living in an environment with a fractal dimension of 1.3 (on the visual plane, that is – the actual physical landscape would have a dimension between 2 and 3). Taylor has suggested that the fractal dimension gives us a quick assessment of an environment's degree of complexity. And the complexity of a savannah offered the best chances for survival because it reduced the chance of failing to see a predator but, at the same time, provided hiding places. A less complex environment like a desert would not provide enough shelter and a dense jungle would not allow us to see an approaching predator in time to flee. This evolutionary explanation could also be the reason why the human brain appears to be optimized to process fractals of a low to intermediate dimension.

So it seems that it is the fractality of structures and processes which is the basis of their aesthetic appeal. If internal and external structures are of the same fractal dimension, they match in terms of their scaling relations. This is not only true for spatial fractals but also for temporal scaling structures such as pink noise. Although pink noise is ubiquitous, we do not know why the physical world changes over time at this distribution. Situated between white noise, which is uncorrelated, and brown noise, which shows stronger correlations, pink noise is a fractal of the intermediate range. Because it can be found both within our bodies and in the external world, pink noise is a promising candidate for temporal matching between internal and external dynamics. Within us, it can be found on almost all levels, from cellular dynamics to our sensory-motor functions. Its prevalence is often explained, in evolutionary terms, as a selection effect, as we have evolved within an environment which abounds with pink noise. As a result, we have probably integrated this pattern into our sensory-motor systems and, possibly, also into the way we store and retrieve memories. The resulting matching between inside and outside very likely works both ways: On the one hand, we perceive what we know, and what we know is what we resonate. On the other hand, the patterns we perceive shape our perceptual apparatus. So in the case of pink noise, we are probably dealing with both an externalization of the interior and an internalization of the exterior.

As pink noise is also a fractal of the intermediate range, Joye has looked into the question as to whether it has a stress-reducing effect on humans, as did the savannah painting. A low to intermediate fractal dimension reflects a degree of complexity between order and chaos, which allow for low-stress interaction with the world – somewhere between boredom and a threatening amount of novelty. An individual exposed to low stress has a wider range of possible behavioural responses, which is of advantage when he has to adapt to changes in his environment. As, among others, Berntson and Cacioppo have pointed out, stress correlates with lower variability [33]. Healthy people have a larger heart rate variability – that is to say, their internal dynamics displays more complexity, which gives them a wider choice of reactions to environmental changes than people who suffer from, say, heart disease or stress. As West has pointed out, a loss of internal complexity reduces variability – in other words, the heart's dynamics no longer displays its fractal dynamics [34]. The same is true for the dynamics of other physiological systems, such as our vascular system and our bronchial tubes. A fractal time series indicates healthy heartbeat, breathing and walking.

Joye extrapolates on these findings and speculates that "by administering fractal patterns or qualities to stressed individuals, efficient functioning, and the associated fractal spectra, can be restored" [35]. He admits that experimental evidence for this hypothesis is not overwhelming and more research is required. However, there are a number of studies, such as Taylor's, which suggest that spatial fractals in the low to intermediate range reduce physiological stress. Joye also refers to experiments conducted by Muzalevskaia et al, in which animals and humans who were exposed to pink noise in the form of a weak magnetic field showed improved immune responses and reduced cancer growth [36]. Bert De Coensel et al has looked into the effect of natural structures, in particular, 1/f noise (pink noise), on people's well-being and concluded that "although a direct relationship between natural soundscapes an psychological restoration has to our knowledge not been proven scientifically till today, the body of indirect evidence of its importance is strong" [37]. Indirect evidence means that humans show an aesthetic preference for natural patterns, which usually exhibit fractal dynamics – in particular pink noise [38, 39, 40]. In earlier work, De Coensel et al built on the now well-established observation that music of very different genres all display 1/f dynamics [41], and looked for 1/f noise in urban soundscapes [42]. They found the expected 1/f behaviour in the frequency interval [0.2Hz, 5Hz], which corresponds to time scales of a few seconds. For longer time intervals, the loudness gradient was steeper than 1/f – which, in analogy with the findings made for music – De Coensel et al interpreted as an indication of too much predictability and, therefore, conducive to boredom. Gradients less steep than 1/f are too unpredictable to be perceived as music-like. The authors cautiously speculate that predictability may be a promising measure of the harmful effect of noise on humans. In a similarly indirect way, this view is supported by Estate Sokhadze's work on the restorative effects of music – which happens to be pink noise – on stressed humans [43].

Internal and external dynamics can lock into each other in a nested or a non-nested manner. The statistically self-similar dynamics of pink noise occurs simultaneously on nested LODs – that is to say, shorter time spans display the same degree of variability as longer, embedding time spans. If their nesting manifests itself as integration, a perceptional gestalt evolves, which is not characterized by congruence, as in the binding-by-synchronization scenario, but by a resonance of scaling dynamics. This nested communication between internal and external dynamics, as in the case of pink noise, creates a more complex gestalt than synchrony, and a different window of the Now. Whereas synchrony is characterized by a loss of Δt_{depth}, nesting generates a fractal temporal perspective and therefore, a more extended Now. This is so because the various LODs involved in the nesting overlap and thus also create Δt_{length} on each LOD. Evidence supporting this hypothesis comes from research into schizophrenia, which shows that our temporal window of multi-sensory binding can be narrowed by perceptual training [44]. The underlying assumption here is, of course, that the temporal window of multi-sensory binding corresponds, on the phenomenological level, to our window of the present – that is to say, our Now. Let us look at an experiment which justifies this connection.

Powers et al [45] conducted an experiment in which schizophrenics and controls were asked to decide whether visual and auditory stimuli (a flash and a beep) were simultaneous or not. The stimuli were connected in such a way that they would somewhat overlap over a range of delays. Participants had to press a button when they thought that the flash and the beep occurred at the same time. Because of the staggered overlapping of almost simultaneous stimuli, the resulting windows, in which the auditory and visual event were merged into one, differed in size. So reports of simultaneity for non-simultaneous conditions would indicate an extension of that individual's temporal window of multi-sensory binding. Vice versa, judging fewer and fewer non-simultaneous events as simultaneous would point to a narrowing of the window. Powers et al's simultaneity judgement test produced a surprising result: After a day of training, there was a significant change in participants' judgement of perceived simultaneity between the auditory and visual stimuli: Those individuals who had to react by pressing a button judged increasingly fewer non-simultaneous events as simultaneous – in other words, their temporal window was narrowed. This change lasted for about a week after cessation of training.

A second experiment ran simultaneously with the one I have described. In it, a group just passively watched the simultaneous and non-simultaneous flashes and beeps. The participant of this group, who were not required to react by pressing a button, surprisingly showed an increase in window size during the course of the training week. This suggests that perception without action does not yield the temporal window. In both experiments, the main modification in window size happened surprisingly fast – that is to say after one day of training, and the effect lasted for approximately a week after cessation of training. Powers et al conclude that the possibility of inducing lasting changes in individuals' multisensory temporal binding windows may provide strategies to deal with certain disorders that are associated with altered temporal integration, such as Parkinson's Disease and schizophrenia. (We shall return to this topic in Chapter 9, in which we shall discuss *temporal misfits*). These findings are of importance to the notion of simultaneity as defined by Δt_{depth}: if we can modify the extension of our Now by training, we can change our temporal perspective and thus create a different reality, with a custom-made causality. This is so because the order of events we create via the determination of succession depends on our judgement of what is simultaneous and what is non-simultaneous.

In the next chapter, we shall investigate how the extension of our window of the Now and our nesting speed determine our perspectives, judgements and moods.

References

1. G. Hüther, Die Macht der inneren Bilder: Wie Visionen das Gehirn, den Menschen und die Welt verändern. Vandenhoeck & Ruprecht, Göttingen, 2004.
2. I.P. Pavlov, Conditioned Reflexes: An Investigation of the Physiological Activity of the Cerebral Cortex. Ed. by G.V. Anrep, Oxford University Press, London 1927.
3. Pestalozzi. Sämtliche Werke. Ed. by A. Buchenau, E. Spranger and H. Stettbacher. Gruyter, Berlin/Zürich 1927-1996.
4. G. Hüther, Wie Embodiment neurobiologisch erklärt werden kann, in: Maja Storch et al: Embodiment. Die Wechselwirkung von Körper und Psyche verstehen und nutzen. Huber, Bern 2006, pp. 92/93.
5. G. Buzsáki, Rhythms of the Brain. Oxford University Press, Oxford, 2006.
6. ibid.
7. ibid, pp. 260-261.
8. W. Singer, Consciousness and Neuronal Synchronization, in: The Neurology of Consciousness, Ed. by S. Laureys and G. Tononi, Elsevier, New York 2009, pp. 45-46.
9. ibid, p. 47.
10. ibid, p. 47.
11. ibid, p. 49.
12. ibid, p. 50.
13. ibid, p. 40.
14. Nano Extra, Die Frequenz der Stille: Forschung zur Meditation, broadcast on 3sat (German TV) on 11.6.2006.
15. M. Kaufman, Meditation Gives Brain a Charge, Study Finds, in: Washington Post, 3 January 2005.
16. Center for Mindfulness in Medicine, Health Care and Society. www.umassmed.edu/content/aspx?id=41252.
17. B.J. West, Where Medicine Went Wrong – Rediscovering the Path to Complexity, World Scientific, Singapore 2006.
18. D.M. Eagleman, Brain Time, www.edge.org/3rdculture/eagleman09/eagleman09index. html, 20.10.2009, p. 5.
19. ibid, p. 5.
20. ibid, p. 6.
21. A. Noë, Action in Perception, MIT Press, Cambridge 2004.
22. ibid, p. 7.
23. D. De Grave, personal communication 2005.
24. L. Nottale and P. Timar, Relativity of Scales: Application to an Endo-Perspective of Temporal Structures, in: Simultaneity – Temporal Structures and Observer Perspectives, Ed. by S. Vrobel, O.E. Rössler, T. Marks-Tarlow, pp. 229-242.

25. D.M. Eagleman, Brain Time, www.edge.org/3rd*culture/eagleman09/eagleman09*index. html, 20.10.2009, p. 1.
26. S. Vrobel, When Time Slows Down: The Joys and Woes of De-Nesting, Talk at InterSymp – International Conference on Cybernetics and Systems Theory, Baden-Baden, Germany 2010.
27. Erimaki, S., Kanatsouli, K., Tsirka, V., Karakonstanaki, E., Sakkalis, V., Vourkas, M. and Micheloyannis, S., EEG Responses to Complex Fractal Stimuli. Poster presentation at 2nd Nonlinear Sciences Conference, Heraklion, Crete, 2006.
28. Illustration courtesy of Rodd Halstead 2010.
29. Y. Joye, A Tentative Argument for the Inclusion of Nature-Based Forms in Architecture. PhD thesis, University of Ghent, Belgium 2007.
30. C.-K. Peng, J.E. Mietus, Y. Liu, J.M. Hausdorff, H.E. Stanley, A.L. Goldberger, L.A. Lipsitz, Quantifying Fractal Dynamics of Human Respiration: Age and Gender Effects, in: Annals of Biomedical Engineering, Vol. 30, No. 5, May 2002, pp. 683-692, Springer 2002.
31. C.M. Caroline, M. Hagerhall, T. Purcell and R. Taylor, Fractal Dimension of Landscape Silhouette Outlines as a Predictor of Landscape Preference, in: Journal of Environmental Psychology, Vol. 24, No. 2, June 2004, pp. 247-255.
32. R.P. Taylor, Reduction of Physiological Stress Using Fractal Art and Architecture, in: Leonardo, Vol. 39, No. 3, June 2006, pp. 245-251.
33. G.G. Berntson and J.J. Cacioppo, Heart Rate Variability: Stress and Psychiatric disorders, in: Dynamic Electrocardiography, Ed. by M. Malik and J.A. Camm, Futura, New York 2004, pp. 56-63.
34. B.J. West, Where Medicine Went Wrong – Rediscovering the Path to Complexity, World Scientific, Singapore 2006.
35. Y. Joye, A Tentative Argument for the Inclusion of Nature-Based Forms in Architecture. PhD thesis, University of Ghent, Belgium 2007, p. 157.
36. N.I. Muzalevskaia and V.M. Uritskii, Antitumor Effect of a Weak Superflow Frequency Stochastic Magnetic Field with 1/f Spectrum, in: Biofizika, Vol. 42, No. 4, July-August 1997, pp. 961-70 (in Russian).
37. B. De Coensel and B. Botteldooren, The Quiet Rural Soundscape and How to Characterize it, in: B. De Coensel, Introducing the Temporal Aspect in Environmental Soundscape Research, Dissertation, University of Ghent 2007, p. 172.
38. T.R. Herzog, C.P. Maguire and M.B. Nebel, Assessing the Restorative Components of Environments, Journal of Environmental Psychology, Vol. 23, No. 2, 2003, pp. 159-170.
39. T. Hartig, M. Mang and G.W. Evans, Restorative Effects of Natural Environment Experiences, in: Environmental Behaviour, Vol. 23, No. 1, 1991, pp. 3-26.
40. H. Staats, A. Kieviet and T. Hartig, Where to Recover from Attentional Fatigue: An Expectancy Value Analysis of Environmental Preference, in: Journal of Environmental Psychology, Vol. 23, No. 2, 2003, pp. 147-157.

41. R.F. Voss, Fractal Music, in: The Science of Fractal Images, Ed. by H.-O. Peitgen and D. Saupe, Springer, London 1988.

42. B. De Coensel, B. Botteldooren and T. De Muer, 1/f Noise in Rural and Urban Soundscapes, in: Acta Acoustica united with Acoustica, Vol. 89, 2003, pp. 287-295.

43. E.M. Sokhadze, Effects of Music on the Recovery of Autonomic and Electrocortical Activity After Stress Induced by Aversive Stimuli, in: Appl. Psychophyiol. Biofeedback, Vol. 32, 2007, pp. 31-50.

44. A.R. Powers III et al, Perceptual Training Narrows the Temporal Window of Multisensory Binding, in: J. Neurosci., 30 September 2009, pp. 12265-12274.

45. ibid.

Chapter 7

Nesting vs Global and Local Perspectives

When neuropsychologist David Eagleman asked people to jump off a 50-metre tower into a net, he was trying to find out whether perceived slowing of time was due to high-speed neural snapshots (as discussed in Chapter 1) [1]. The idea was to see if the slowing down of time in stressful situations, such as an adrenaline-spilling free fall, results from higher temporal resolution. However, the jumpers, who were instructed to stare at flashing numbers on displays strapped to their wrists, did not perceive the numbers at an increased temporal resolution during their jump. Eagleman concluded that we can rule out a higher temporal resolution as the cause of time slowing down. He suggests that subjective duration is linked to the amount of detail we process – that is to say, if time, in retrospect, appears to have slowed down during a stressful situation, this is so because more memories are laid down during such situations.

Temporal resolution, however, extends not in the dimension of Δt_{length} – that is to say, a succession of oscillations on one LOD – but in the dimension of Δt_{depth}. Eagleman's experiment looked at Δt_{length} only – successive numbers flashing slightly faster than can normally be perceived. The temporal resolution of Δt_{depth} is determined by the number of simultaneous LODs perceived – in other words, by the number of nestings. In this chapter, I shall differentiate between nested, local and global perspectives and suggest that a slowing down of time is due to a reduction of Δt_{depth} [2].

A reduction of Δt_{depth} results from de-nesting, when we do not take account of an embedding context (I use the terms de-nesting and de-contextualization synonymously) or from directing attention to a specific

139

LOD. De-nesting often occurs spontaneously, as was the case with the police officer in Chapter 1 who experienced a slowing down of time. It can also be trained, as in deep meditation, when we rid ourselves of all contexts and focus only one LOD – say, a mantra.

Contexts can suddenly appear or vanish, but can also vary by degrees, for instance, when luminance is slightly reduced or enhanced. Surprisingly, the amount of contrast we perceive also determines the speed at which we perceive embedded objects. Stuart Anstis has shown that moving objects appear to slow down when they enter a low contrast environment [3, 4]. This would explain why we misjudge our own speed or that of other cars in fog: Anstis explains such misjudgement with the fact that when objects are in low contrast, as they are in a fog, they appear to move more slowly:

> In a fog, other cars are reduced in contrast so they appear to be going more slowly than they really are. Also, a driver judges his own speed largely by visual cues from the landscape as it slides past him, often viewed through the side windows of the car in peripheral vision. (...) Fog reduces the contrast of the passing landscape, so it appears to slip by him more slowly and he believes that he himself is driving slowly [5].

In an experiment, Anstis has shown that apparent speed varies with contrast [6]. He moved two squares, one light and one dark grey, at constant speed over a background of vertical stripes (see Fig. 7.1). The squares were slightly wider than the stripes, so their front and back edges always covered the same background colour (a white or a black stripe). Anstis observed that the two squares seemed to stop and start in alternation, like a pair of walking feet – a perceptual distortion he called the "footstep illusion".

> When the dark grey square lay on white stripes, it had high contrast (dark versus white) and appeared to speed up momentarily. When it lay on black stripes it had low contrast

(dark versus black) and appeared to slow down. The opposite was true for the light grey square. [7]

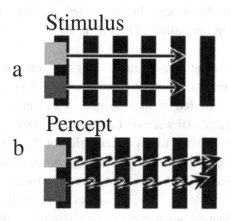

Fig. 7.1. The footstep Illusion. Reprinted from S. Anstis, Moving objects appear to slow down at low contrast, in: *Neural Networks* 16 (2003), p. 934, with permission from Elsevier.

The temporal distortion arose from the change in contrast of the embedding background and the nested square. Unlike the still embedding and nested disks in Dakin's experiment with schizophrenics (see Chapter 3), the simultaneous contrast in Anstis' experiment was changing at a continuous rate. So the temporal illusion of an embedded object speeding up or slowing down was brought about by a changing simultaneous context. To be precise, it was the difference in luminance between dark grey and light grey which increased or reduced the simultaneous contrast between nested and embedding objects. Had both greys been of the same scale, the embedded object would obviously have become invisible, just like a grey car in very dense fog. However, as soon as there is enough contrast for us to perceive the "difference that makes a difference" (Bateson' famous definition of information [8]), a change from a higher to a lower contrast will result in a partial de-contextualization. Vice versa, the change from a lower contrast to a higher one leads to partial contextualization. Our fractal temporal perspective thus changes with the degree to which a background is modified: We interpret an embedded object as nested only if the simultaneous contrast is strong enough to make a difference to us. In other words, no Δt_{depth} can be generated if there is no perceived difference between the nested LODs, as is the case when we try to make out the grey car in fog.

Anstis' experiment shows that a higher contrast generates more context (which results in more simultaneous LODs), which, in turn, speeds up objects nested in that context. Vice versa, a lower contrast creates less context (fewer simultaneous LODs) and slows down embedded objects. That is to say: a reduction of Δt_{depth} slows down time, an increase in Δt_{depth} speeds it up. This interpretation is consistent with the subjective distortions of duration I have described earlier.

A reduction of Δt_{depth} can be achieved by ridding ourselves of contexts like unwanted onion peels or by reducing the degree of contrast. It may also be obtained by directing our attention to one LOD only. The focus of attention may be shifted to a nested object or to an embedding context in the shape of a global structure. We tend to group visual percepts, such as the motion of individual dots, possibly in order to lower the level of complexity of the moving image. Such de-complexification would very likely be a selection effect, as it allows us to navigate the world faster and to keep track of meaningful ensembles rather than getting lost in insignificant detail. However, as we have seen in Chapter 3, it can also be of advantage to see the trees instead of the forest, as this enables us to compare and match patterns without experiencing an unwelcome simultaneous contrast which distorts our perception of the nested structure.

In a separate experiment, Anstis has shown that, when we look at four pairs of dots which rotate at constant speed (i.e. 1 rev/sec.) around their pair centres, we first perceive the local motions – that is, the clockwise rotation of each pair. But soon, we perceive two global structures instead of the pairs of dots: two overlapping squares appear, which slide over each other (see Fig. 7.2). When we shift our visual focus, we start out with the local perspective again:

The display tended to flip back and forth over time between local and global motion, although the physical display never alters. In other words the ambiguity in this display lay not in the motions themselves, but in the perceptual groupings, or solutions that the visual system adopted to the 'binding problem'. [9]

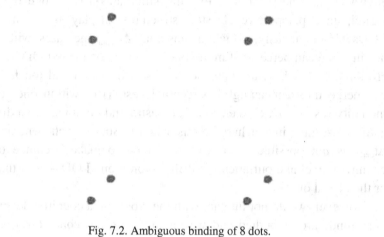

Fig. 7.2. Ambiguous binding of 8 dots.

Anstis points out that global and local perspectives are incompatible solutions to the binding problem because it is not possible to see both movements simultaneously. He suggests that local motion is pre-attentive and global motion is attentive.

The experiment was further examined and extended by Peter Kohler et al, who discovered that when our perspective changes from local to global, the global configuration appears to move more slowly than the local one [10]. To experience the phenomenon, I recommend visiting Kohler's animations which can be accessed online [11]. Although numerous attempts have been made to explain why we perceive the global-motion percept as moving more slowly than the local ones, Kohler states that it remains unclear.

When we look at two revolving squares rather than four pairs of dots, we reduce the level of complexity of the percept, because we form a gestalt and simplify the world by shifting from a local to a global perspective.

It is important to keep in mind that, on a cognitive level, global and local perspectives are temporally incompatible. In other words, we see either the forest or the trees, but never both simultaneously. However, from the embodied point of view, it is quite possible to envisage multi-nested LODs existing simultaneously in one observer-participant's perspective. Nested structures display their embedded LODs simultaneously, which means that Δt_{depth} increases with every nesting. We can perceive stimuli simultaneously on many LODs, be they cross-modal (such as the perception of sound, vision and touch, which we merge into a meaningful perceptual gestalt) or within one sensory modality (as when we listen to an orchestra and combine the individual auditory stimuli into a harmonious whole). Such a rich generation of Δt_{depth} is not possible when we are forced to make a choice on the cognitive level and our attention drifts to only one LOD – be it the local or the global one.

However, we do not interact with the world on a cognitive level only. Our minds are embodied and we interact with our context via sensory channels simultaneously. Our fractal physiology allows both for 'vertical communication' within our bodies – that is to say, when we integrate rhythms or fluctuations across levels and time scales [12] and for multi-levelled interaction with our environment. Our nested rhythms can entrain with their context on a number of LODs simultaneously. Thus, our bodies have no problem reconciling a global with a local perspective and actually experiencing both simultaneously: We may be entrained to the circadian rhythm (which is a safe time-setter that regulates our sleep and wake cycles with all associated changes in, for example, body temperature and metabolism) and, at the same time, lock into a meditation stimulation at 10 Hz. This, though, may trigger a panic attack (because 10 Hz stimuli often produce 30 Hz brain waves: see Chapter 2). Simultaneous, nested stimuli, be they in one sensory mode, as when we listen to an orchestra, or cross-modal, when we hear, see and smell a fireworks display with musical accompaniment. Multi-levelled entrainment is our primary mode of interaction with the world. A limitation exists only on the cognitive level, when we have to make a choice between the forest and the trees.

Fig. 7.3. An artist's impression of a reconciled global and local perspective. [13]

The notions of global and local perception are related to the cognitive styles of field-dependency and field-independency which American psychologist Herman Witkin developed in the 1950s [14]. In a nutshell, field-independent people see the forest, while field-dependent ones see the trees. This second group has difficulty identifying embedded shapes if there is little contrast between the nested objects and the embedding background. They also blend in better and work in teams more easily, whereas field-independent individuals effortlessly identify nestings and tend to make decisions alone. There appears to be an – at least indirect – correlation between a person's cognitive style and his ability or willingness to identify and set boundaries between nested objects and their embedding context or between self and non-self. Thus, we might encounter strong field-dependence in the boundary syndrome, when individuals fear "losing or fusing" (see Chapter 9) or in a meditating monk, who is at one with the universe and in a state of total immersion, and for whom boundaries are dissolved. In contrast to the global/local distinction, field-independence and field-dependence imply nesting, as the individual either embeds himself into his context or de-nests – that is to say, isolates himself from his environment, in order to fully develop his potential.

A speeding up or slowing down in time has often been correlated with moods: Time flies when we're happy and drags when we're sad. This correlation may possibly also be described by an increase or a reduction of Δt_{depth}. We may look here at an experiment conducted by Emily Pronin and Daniel Wegner, which showed that thought speed influences moods [15]. The experiment revealed that fast reading improved people's moods, irrespective of positive or negative content. Moreover, when people felt they were reading fast, they also reported feelings of power and creativity, a heightened sense of energy and inflated self-esteem, which are often associated with mania. Pronin and Wegner looked at the effects not only of fast, but also of slow thinking:

> (...) if accelerated thinking leads to experiences associated with mania, might decelerated thinking lead to experiences of depression? A brief examination of the criteria for a depressive episode supports the idea of a link between depression and decelerated thinking: Whereas racing thoughts are a criterion for diagnosing a manic episode, 'diminished ability to think' is a criterion for a depressive episode. (...) The results of our experiment suggest the intriguing possibility that even during moments when we feel stuck having depressed thoughts, interventions that accelerate the speed of such thoughts may serve to boost feelings of positive affect and energy. [16, 17]

Although participants were reading aloud statements which appeared letter by letter, they were able to integrate the individual words into a semantically meaningful whole. This cognitive task of grasping the meaning of a sentence involves nesting, as we integrate short phonetic units into a semantically complete gestalt (syllables are nested into words, words into sentences and sentences into larger semantic contexts). Therefore, I am inclined to interpret the performances in Pronin and Wegner's experiment as fast nesting: equating thought speed with nesting would mean that happy moods arise from fast nesting. While further research is needed to determine whether thought speed is a result of fast nesting, I may suggest two experiments which would assist in this endeavour:

1. Participants could be asked quickly to read a text consisting of nonsense words which cannot be contextualized semantically, so that no new meaning and, therefore, no Δt_{depth} would be generated. Neither should the text have a grammatical structure which may trigger nesting activities. If the participants' moods do not improve during the fast reading of the nonsense text, this would support my suggestion that it is not an unspecific processing speed, but nesting speed – that is to say, a fast increase of Δt_{depth} – which generates a positive mood [18].

2. Participants could be exposed to both nested and non-nested visual stimuli. Zooming out of Google Earth would be an example of a nested stimulus, as every zooming step embeds the previous image into a wider context. A visually non-nested stimulus would move from one town to another on the same scale. The presentations should cover the same externally measured interval in Δt_{length}. If participants navigating on one scale (LOD) experienced a longer duration than those zooming out via a number of nested scales, this would suggest a correlation between a non-nested perspective and the experience of time slowing down. Depending on the selected control group, the opposite correlation also holds: If participants who zoom out perceive a speeding up of time – that is to say, a shorter relative duration of the unit interval – this would suggest a correlation between a nested perspective and the experience of time speeding up. In other words, zooming out generates Δt_{depth} as a result of nesting and speeds up time (i.e. it decreases subjective duration). Remaining on one LOD does not generate Δt_{depth} and slows down time (i.e. it increases subjective duration) [19].

If it is indeed our nesting speed which determines our mood and people are primed into a positive or negative mood by fast contextualization, this would have far-reaching consequences because our moods determine our judgement and whether we take on a global or local perspective [20]. We shall look more closely into these correlations in Chapter 9. For now, it shall suffice to acknowledge that processing speed is linked to moods in such a way that accelerated speed correlates with positive mood (and other manifestations of mania).

An interesting quirk was given to the issue of mood and speed by Aaron Sackett et al, who established a correlation between the extent to which we enjoy a task and distortions in subjective duration. Apparently, there is a direct link between our temporal modality – that is to say, our feeling of how slowly or quickly time passes – and the degree to which we enjoy a task. It seems to be common knowledge that time passes more quickly when we enjoy ourselves and more slowly when we are suffering. But although there has always been plenty of anecdotal evidence, it is only recently that we know this correlation works both ways: Not only does time fly when we enjoy ourselves, but we also judge a task as having been more enjoyable if we believe that we underestimated its duration. Vice-versa, if we think that we have overestimated the time period, we tend to judge the task as having been less enjoyable. In Sackett et al's experiment, people were induced to believe that time passed more quickly than they had expected [21]. The perceived time distortion led them to judge tasks and songs as more enjoyable and noises to be less annoying – possibly because they were looking for an explanation for the temporal distortion and came up with a convincing answer:

> The feeling that time has been distorted may lead people to question why. We propose that, to make sense of such a distortion, people infer that subjective time perception carries information about the hedonic value of the experience at hand. Specifically, people may rely upon a naive theory that enjoyment accelerates the passage of time – while suffering decelerates it. We hypothesize that if people think time has passed unusually quickly (slowly), they will evaluate the experience or activity that filled that time more positively (negatively). [22]

This hypothesis was tested by misinforming people about the time a specific task had taken – the task being to underline double-letter combinations in a text. Participants were told it would take about ten minutes. However, after a surreptitious exchange of stopwatches, the experimenter returned to the room after 5 minutes, pretending 10 had passed. As expected, the participants were surprised at how fast time had flown. They were asked to rate the task in terms of enjoyment and also judge how quickly time had seemed to pass. In a sister-experiment, which induced the 'time drags' condition (that is to say, the experimenter re-entered the room after 20 minutes and announced that 10 minutes had passed), participants also evaluated the enjoyment and temporal distortion of the task. Here, too, the results matched Sackett's expectation:

> As hypothesized, participants in the time-flies conditions rated the task as more enjoyable than did those in the time-drags conditions. (...) These results suggest that people evaluate tasks more favourably when time seems to have passed surprisingly quickly rather than surprisingly slowly. [23]

Of particular interest to us, though, is why our perception of a temporal distortion determines how enjoyable or annoying we rate a paticular task or noise level. We seem to be conditioned to such an extent that the perception of temporal distortions is hard-wired to our estimation of the level of enjoyment or annoyance. And it appears that the neural correlate of the experience of hedonism fires off a whole neural assembly, which would also happen to fire together if we experienced time as speeded up. Is that all there is to it? It is also conceivable that an increase in Δt_{depth} arouses the neural assembly.

When the formation of Δt_{depth} is compromised, mood disorders are often the first indicators of a pathological temporal perspective. An example is depression, which often leads to a mismatch of retension and protension (memory and anticipation) in the Now: A strong decrease in protension effectively reduces the number of nestings in the Now, as past and current events are not nested into anticipated ones. In other words, a reduction of Δt_{depth} leads to depression if that depth solely consists of a person's nested memories and lacks the other half, namely, anticipation. Emrich et al have shown that, in depression, an individuals' experience is dominated by the past [24, 25]. They carried out an experiment in which two groups of participants – depressive individuals and a control group – watched a video monitor on which a series of words with different emotional connotations was shown. The corresponding event-related brain potentials (ERPs) were recorded. The two groups had to rate every word presented as new or as having been presented before. There were great differences between the groups' responses to old and new words with emotional content ('old' meaning that the word was not perceived as new). While the ERPs of participants in the control group showed significant differences when exposed to new and old words, in ERPs of depressive participants, such differences were hardly detectable. Emrich et al concluded that negative cognitions formed the expectations of the depressed individuals: they expected negative words much more than positive or neutral ones (negative memories, in particular, led to a dramatic reduction of old/new differences). The dominance of retention and the lack of protension compromised the depressed individuals' ability to integrate negative items into a new positive or neutral context. In terms of fractal time, one may say that their ability to generate Δt_{depth} was incapacitated.

Emrich et al's findings suggest that depressive patients had lost their ability to embed their (negative) past retensions and protensions into positive new contexts. However, this does not mean that they were incapable of anticipation. They simply filtered their anticipatory perspective in such a way that it would give attention almost exclusively to negative stimuli and blot out positive ones: the depressed participants' ability to perceive a positive new stimulus was highly compromised. On the other hand, they did nest their past into stimuli with negative connotations (which their protension was conditioned to anticipate). If we draw primarily on the past in our Nows, we generate a temporal perspective which is dominated by retensions. This mechanism has a conditioning effect: the longer it persists, the deeper the nesting cascade of retension-heavy Nows grows. Our Now's anticipatory faculty becomes severely narrowed, in the wake of which we miss many a nesting opportunity, because we simply disregard what does not match our (negative) expectations. As a result, depressed individuals cannot perform the (positive) contextualization necessary to change the structure of their Now.

If our ability to generate Δt_{depth} is compromised, time subjectively slows down for us, as new stimuli are arranged on existing LODs. We do not generate new LODs because we do not embed the past into new contexts. Rather than forming new LODs and thus Δt_{depth}, we arrange unanticipated stimuli on existing LODs. This leads to an increase in Δt_{length}, which dilates time – that is to say, it increases duration [26]. The resulting observer-participant perspective is crippled by a narrowed anticipatory faculty. It is a perspective which avoids contextualization. The disadvantages of a retension-heavy Now were already known to Lewis Carroll, whose White Queen says to Alice: "It's a poor sort of memory that only works backwards" [27].

The link between depression and the subjective slowing down of time has been well established [28]. We may interpret this correlation as the result of depressed individuals' compromised ability to generate Δt_{depth}. We shall investigate the pathology of temporal misfits in more detail in Chapter 9.

I have suggested that a slowing down of time is due to a reduction of Δt_{depth}. Further, I have pointed out correlations between our ability to make out a contrast, our grouping and nesting capacity, types of perspective (nested, global, local), time perception, mood and anticipatory faculty (see Table 7.1, below). However, the question remains as to the nature of the causation which generates nestings in the first place. My Theory of Fractal Time assumes that past nestings are embedded in current and anticipated ones, and the structure of our current Now determines how we perceive the world [29]. We saw, for instance, that a retension-heavy Now can distort expectations and, as a result, our perception of the world. If we think of nestings as hierarchies, we may be tempted to consider downward causation as the underlying causation in a fractal temporal perspective. However, this term requires disambiguating, as Menno Hulswit has pointed out [30]. Huswit differentiates between non-circular and circular causality. Non-circular causality describes, for instance, the relation between a virus and the disease it causes in human beings, while circular causality occurs in emergent phenomena such as crowd behaviour:

> Thus, by saying that some rhinovirus explains or conditions or causes the common cold, we mean that the virus explains or conditions or causes the common cold, and that the common cold does not cause the rhinovirus. But when we say that the crowd causes individuals to behave in certain ways, we do say that (a) the behaviour of the individuals in some way causes the behaviour of the crowd and (b) the crowd causes the behaviour of the individuals. [31]

A nested system, such as individuals in a crowd or biochemical processes in a cell, needs to be described in terms of a part-whole relation. Hulswit suggests that part-whole relations are not determined by efficient causation, because this form of causation is a process in time, unlike the synchronic downward causation at work in part-whole relations:

> (...) biochemical processes do not efficiently cause the cell, nor does the cell as such efficiently cause the biochemical reactions of its molecules. The mistake consists in presupposing that the cell and biochemical processes can exist independently from one another and that there can be a causal relationship between them. The biochemical processes do not cause but *constitute* the cell. Contrary to efficient causation, which is a process in time, this is not a process in time, and therefore cannot be efficient causation. [32]

When Hulswit refers to processes in time, he is talking about Δt_{length} only. Because he does not assume Δt_{depth} to be a temporal dimension, he considers the term downward causation to be ill-chosen, as it does not imply causation at all, and should therefore rather be denoted as downward explanation or downward determination. However, he is aware of the problem arising from our central belief that the cause precedes the effect:

> (...) an effective approach to the problem of downward causation requires a radical break with the Western 'substance addiction' in favour of an ontological framework based on the primacy of process and event. [33]

Hulswit has made important headway by distinguishing between various types of downward causation, in particular, the difference between 'bringing about' and 'explaining'. When we look at nested systems, higher-order properties such as embedding rhythms can have a causal effect, for instance, when neural oscillations are embedded in slower metabolic rhythms. Nested rhythms within our bodies display a simultaneous causal relation, because the body is one systemic whole. And, as we have seen in Chapter 4, the boundaries between systemic wholes are highly negotiable. If a neural implant or a walking stick is incorporated, a new systemic whole is formed. Its newly added part can no longer be described meaningfully as being independent of the whole (if the brain has mapped far space to near space and now regards the added part as an extra limb). Of course, if that added part were isolated, it would no longer function as part of the systemic whole.

A temporal fractal is a nested structure in which the nested and embedding parts form one systemic whole (unlike the Russian Matryoshka dolls, which can be pulled out and lined up side-by-side without changing each other's structure). The causation at work in the dimension of Δt_{depth} does not involve succession – it is timeless in Δt_{length} – and is therefore not subjected to an efficient causation in which the cause precedes its effect. The Now, with its nestings of protension and retension, requires both efficient and final causation to be taken into account, as our memory and anticipation determine our present perspective. We almost never live in the present (unless we approach this state by means of deep meditation): Because we have been conditioned to so many delays we are no longer aware of, all our actions are anticipative – in other words, we live in the future. And the outermost nesting of our temporal fractal perspective happens to include just those delays. Thus, Δt_{depth} is based on both efficient and final causation.

In this chapter, I have differentiated between nested, local and global perspectives and have suggested that a slowing down of time is due to a reduction of Δt_{depth}. In the next Chapter, we will further investigate the role of Δt_{depth} when we look at the notions of content and context and how our ability to focus influences our perception of time.

Table 7.1. Correlations between Δt_{depth} and temporal distortions, perspectives, moods, causation and anticipatory faculty.

	High nesting capacity (the generation of Δt_{depth} is facilitated)	Low nesting capacity (the generation of Δt_{depth} is compromised)
Temporal dimension generated	simultaneity: Δt_{depth}	succession: Δt_{length}
Compatibility between LODs	compatible	incompatible
Contrast	high contrast	low contrast
Grouping	no grouping: high complexity	grouping: low complexity
Perspective	nested	global or local
Perceived speed	time speeds up	time slows down
Mood	positive	negative
Anticipatory faculty	enhanced	compromised
Causation	efficient and final causation	efficient causation

References

1. The Eagleman Lab (http://neuro.bcm.edu/eagleman/time.html)
2. S. Vrobel, When Time Slows Down: The Joys and Woes of De-Nesting. *Proceedings of the 22nd International Conference on Systems Research, Informatics and Cybernetics* (InterSymp 2010), Baden-Baden 2010 (forthcoming).
3. S.M. Anstis, Moving objects appear to slow down at low contrast, in: *Neural Networks* 16 (2003), Elsevier Science, pp. 933-938.

4. S.M. Anstis, Factors affecting footsteps: contrast can change the apparent speed, amplitude and direction of motion, in: *Vision Research* 44 (2004), Elsevier, pp. 2171-2178.

5. S.M. Anstis, Moving objects appear to slow down at low contrast, in: *Neural Networks* 16 (2003), Elsevier Science, p. 934.

6. ibid.

7. ibid, 934.

8. G. Bateson, *Steps to an Ecology of Mind*, Ballantine, New York 1972.

9. S.M. Anstis, Levels of Motion Perception, in: Harris, L. and M. Jenkin (Eds), *Levels of Perception*, New York, Springer 2003, p. 99.

10. P.J. Kohler, G.P. Caplovitz and P.U. Tse, The whole moves less than the spin of its parts, in: *Attention, Perception & Psychophysics*, Vol 71, No. 4, pp. 675-679 (2009), The Psychonomic Society, Inc.

11. P.J. Kohler, *Local Versus Global Motion* (animation 2009). Link http://vimeo.com/3651331

12. C.M. Anderson, From Molecules to Mindfulness: How Vertically Convergent Fractal Time Fluctuations Unify Cognition and Emotion, in: *Consciousness & Emotion* Vol. 1, No. 2 (2000), John Benjamins Publ. Co., pp. 193-226.

13. Photograph courtesy of Julie Neumann 2009 (www.julieneumann.com)

14. H.A. Witkin, C.A. Moore, D.R. Goodenough, P.W. Cox, Field-dependent and Field-independent Cognitive Styles and their Educational Implications, in: *Review of Educational Research*, Vo. 47, pp. 1-63, 1977.

15. E. Pronin and D.M. Wegner, Manic Thinking. Independent Effects on Thought Speed and Thought Content on Mood. *Psychological Science* 17 (9) (2006), pp. 807-813.

16. ibid, p. 809.

17. ibid, p. 810.

18. S. Vrobel, Time Slows Down in Nows Deprived of Their Anticipatory Faculty, in: D.M. Dubois (Ed.), *Intl. Journal of Computing Anticipatory Systems*, Vo. 21 (2008), CHAOS, Liège, pp. 85-96.

19. ibid.

20. K. Gasper and G.L. Clore, Attending to the Big Picture: Mood and Global Versus Local Processing of Visual Information. *Psychological Science*, Vol. 13 (2002), No. 1, pp. 34-40.

21. A.M. Sackett, T. Meyvis, L.D. Nelson, B.A. Converse and A.L. Sackett (2010), You're Having Fun When Time Flies: The Hedonic Consequences of Subjective Time Progression, in: *Psychological Science*, Vo. 21, No. 1 (2010), pp. 111-117.

22. ibid, pp. 112ff.

23. ibid, p. 114.

24. H.M. Emrich and D.E. Dietrich, On Time Experience in Depression – Dominance of the Past, in: Thomas Schramme and Johannes Thome (Eds.) *Philosophy And Psychiatry*. http://books.google.com/, 2005, pp. 242-256.

25. H.M. Emrich, C. Bonnemann and D.E. Dietrich, On Time Experience in Depression, in: S. Vrobel, O.E. Rössler and T. Marks-Tarlow (Eds.) *Simultaneity – Temporal Structures and Observer Perspectives*. World Scientific, Singapore 2008.

26. S. Vrobel, *Fractal Time*. The Institute for Advanced Interdisciplinary Research. Houston 1998.

27. L. Carroll, *Through the Looking Glass* (1871). Penguin Popular Classics, London 1994, p. 80.

28. Cf. (for example): S. Gil and S. Droit-Volet, Time Perception, Depression and Sadness, in: *Behavioural Processes*, Vol. 80, No. 2 (2009), pp. 169-176.

29. S. Vrobel, *Fractal Time*. The Institute for Advanced Interdisciplinary Research. Houston 1998.

30. M. Hulswit, How Causal is Downward Causation? in: *Journal for General Philosophy of Science*, Springer, Vol. 36 (2006), pp. 261-287.

31. ibid, p. 265.

32. ibid, p. 274.

33. ibid, p. 284.

Chapter 8

Duration: Distributing Content and Context

17 April brought the first sunny Spring day of 2010 to the German town of Bad Nauheim. My Saturday walk was most pleasant, yet something seemed wrong. Everything appeared calmer, people seemed to be taking more time chatting to each other or crossing the street. The town had a pace I remember from the mid-1960s, when I would feed the ducks in the local park or promenade down Parkstrasse with an ice cream. I was puzzled and it was only when I stopped to look up to the sky that I realized what was different: I was walking – for the first time in my life – through my home town under a crystal-clear blue sky unscathed by vapour trails. Next, I noticed the absence of the faint noise of aircraft taking off from and descending into Frankfurt. It then occured to me that I had never consciously registered the weak acoustic background of the jets. But on those days, when almost all of European airspace was closed due to an eruption of the Icelandic volcano Eyjafjallajökull, I was experiencing the *absence* of jet noise.

When context, such as this weak background sound, suddenly disappears, one's focus is narrowed, albeit unconsciously. And although my memory of that Spring morning serves at best as merely anecdotal evidence, there is – as we saw, for instance, in Chapter 1 – a correlation between a narrowed perspective and subjective time dilation. In this Chapter, we shall see, through various experiments, how diminished context can make time pass more slowly. We narrow our perspective when we attend to or focus on a certain content and disregard others. The disregarded contents then no longer function as context for the focussed content. Taking away content which served as context is an act of de-nesting (or de-contextualizing), which correlates with subjective time dilation – that is to say, time slows down from a particular perspective. This nested (fractal) temporal perspective is shaped by the number of embedded LODs – that is to say, Δt_{depth}. I am suggesting that subjective duration is determined by the way percepts are internally arranged – either as content in the dimension of Δt_{length} or as context in the dimension of Δt_{depth}.

In Chapter 5, I defined context as simultaneous structures in one temporal nested perspective. We arrange these simultaneous structures in such a way that embedded context is interpreted as content. Vice-versa, content may also take the role of context if it functions as an embedding structure itself. But what does it actually mean to think of content as that which is embedded or contained within a wider context? To contain something means to hold it together (Lat. *con* = together, *tenere* = hold). What, then, is the container which holds our percepts together? The context itself may shape a nested event, may cause a simultaneous contrast, which distorts the embedded process, but it is not a container. The container is our nested (fractal) interface – our Now. To answer the question of how an interface is held together, we need to look at the assignment conditions – that is to say, our rules which decide how we temporally arrange events we encounter. There are two options: We can either arrange percepts successively on one LOD and thereby create the dimension of succession, Δt_{length}, or we can nest them, so they overlap to some extent and thus create the dimension of simultaneity, Δt_{depth} (see Fig. 8.1).

It is not the actual duration of events at any given LOD which determines whether they are nested within or embed a simultaneous event, but rather our focus of attention. A shift in attention can turn content into context – that is to say, foreground into background – or vice-versa. This is true for both spatial and temporal structures. A familiar example of foreground-background illusions is Rubin's vase, developed around 1915 by Danish psychologist Edgar Rubin (see Fig. 8.2). Our brain usually interprets the nested structure as foreground and the embedding environment as background, thereby creating spatial depth. In the case of Rubin's vase, there are two possible interpretations, depending on which section we focus.

A temporal example of a content-context transition is when we try to time an event to determine its duration. We then shift our attention away from the event to be timed in order to focus on the timing task itself. The event is shifted from the centre of attention to the periphery – it turns into the context of timing task, which has now taken the role of content.

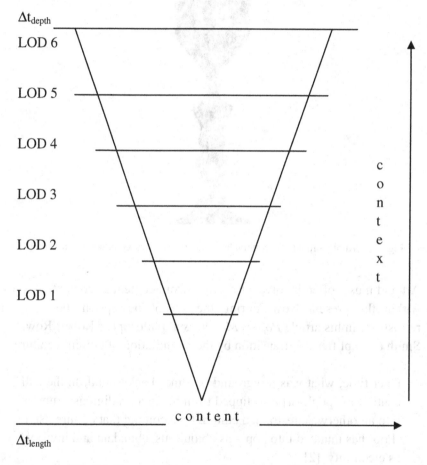

Fig. 8.1. The nested Now – our fractal temporal perspective.

Another example is the auditory illusion of the Shepard scale, described in Chapter 3. Here, we have the choice of focussing either on pitch or on the recurring periodic signal – in other words, we cannot focus on both at the same time. As a result, we interpret either pitch or the periodic signal as content and the other as context.

Fig. 8.2. "Rubin's illusion": Ambivalent foreground – background assignment. [1]

Art and music often involve a transition from content to context not only within the present Now, during the act of perception, but also in retrospect. In his article *Pop as Modernism*, philosopher Robert Rowland Smith exemplifies this transition by the assimilation of pop into culture:

> Over time, what was foreground becomes background. In the half century of 'pop', [it] has slipped from being a prodigious irruption into an otherwise restrained culture to becoming that culture itself. 'Pop' has mutated into pop – as ubiquitous, abundant and invisible as electricity. [2]

He makes the point that the transformation of pop into background means we no longer focus our attention on it – in fact, we assimilate and, thereby, make it invisible:

So 'popular' that it's universal, pop, like weather, is simply there. And yet this background-becoming is exactly what has allowed pop to insinuate itself into our nearest, our innermost worlds. Its ubiquity translates as ignorability, and ignorability makes the perfect guise for penetration; in getting absorbed into the fibres of culture, pop had done a deep dive into the mind. [3]

Interestingly, Smith also remarks that this context works as a vehicle for memory retrieval, as we "recollect a whole summer from a single song, or an entire relationship from an album." [4] A relatively short event like a pop song can trigger within the Now a fractal unfolding of memories which were laid down over months or years. The nested retensions and protensions form an intrinsic context which selects and shapes our current perceptions.

The pop culture example shows that when content becomes context, we give it less attention. This is an important observation, as attention has a direct influence on our perception of duration. We give more attention to new or unexpected events than to familiar and expected ones. So how do our attention and focussing relate to our perception of time and, in particular, duration?

We are all familiar with the 'stopped clock' illusion – the apparent dilation of time we experience when we look at the ticking minute hand of a clock and it seems to be stuck. But time is dilated only during the first tick – after that, it keeps ticking in the expected, regular fashion. What happens in such a situation is that we switch our focus of attention, say, from reading a timetable or heaving suitcases onto the platform, to the platform clock, to check if we can make our connection. This 'chronostasis illusion' – the slowing down of time caused by, for instance, a 'stuck' minute hand – has been investigated by Kielan Yarrow [5]. In Yarrow's experiment, participants had to gaze at a cross on the side of a computer screen, then look at a counter on the opposite side of the screen. The shift in focus from the cross to the counter was accompanied by the corresponding rapid eye movement, known as a saccade. It turned out that participants overestimated the first interval which followed the saccade. Moreover, the longer the saccade took – that is to say, the greater its angle – the greater was participants' overestimation of the duration of the interval.

> (...) the size of the chronostasis effect had grown linearly with the duration of the preceding saccade. It was as if the extra time taken to complete the larger saccade had been added on to the estimated duration (...) [6]

This overestimation of duration of stimuli seen immediately after a saccade links chronostasis with attentional shifts. There appears, then, to be a correlation between selective attention and subjective slowing down of time.

But what actually happens when we shift our attention and focus – in other words, when we de-contextualize and concentrate on one or few LODs? According to the standard attentional model of time perception, if more attention is given to the duration of an event, its perceived duration increases and, vice versa, with less attention to the duration, its perceived duration decreases. Focussing on the timing task only narrows our perspective and decreases Δt_{depth}. De-focussing – that is to say, simultaneous action-perception on various LODs – increases Δt_{depth}. As Tse puts it:

> Duration estimates (...) follow opposite trends in prospective experiments that involve concurrent processing and those that do not. In the absence of concurrent processing, subjective time expands whereas in its presence, it typically contracts. [7]

In terms of fractal time, "concurrent processing" means that simultaneity is generated, as attention is given to both the event and the timing of that event in a parallel manner. In the absence of such simultaneity, time slows down, because events are arranged successively in the dimension of Δt_{length} rather than in Δt_{depth}. If simultaneity is present, however, subjective time contracts because events can now be arranged 'vertically', in the dimension of Δt_{depth}. But what does it actually mean to give attention to duration? A commonly held assumption is that there is a 'counter' – an internal clock – that registers the number of units of temporal information we process while experiencing an event. In Tse's words:

> (...) the number of units of temporal information that are counted decreases when attention is distracted from processing the duration of an interval (...) attention increases duration judgements when duration *per se* is attended because fewer temporal cues are missed. [8]

A fractal interpretation would not state that duration judgements increase because we perceive *more* temporal cues when we focus on a timer. Rather, it is because we focus on duration *per se*, because we arrange percepts on that one LOD which is the centre of our attention – the LOD of the timer – that duration judgements increase. Thus, we generate succession on one LOD, which means that Δt_{length} expands and Δt_{depth} contracts.

Eagleman has shown with his tower-jumping experiment (see Chapter 1) [9] that we cannot explain time dilation with an increased temporal resolution. If subjective time expands during a highly stressful situation such as a free fall, according to Eagleman, it is not due to the 'counter' counting more units at that moment. However, he sticks with the idea that the number of units of information processed is the basis of subjective duration – albeit in retrospect.

I believe that subjective time dilation during attention or stress can be explained by interpreting the temporal distortion as a result of a reduction of Δt_{depth}. Giving attention to one aspect of an event and excluding others means that we give that privileged aspect the status of content (rather than context). This narrows our temporal perspective and, thereby, decreases Δt_{depth}. Of course, this content is an LOD which may, just like any other, serve as context to an embedded content in the shape of a time setter. This is possible because LODs are defined as levels whose respective successive temporal intervals are of equal extension.

Tse has conducted a series of experiments which support the standard attentional model [10]. However, they revealed a threshold below which no time dilation was perceived. Participants watched a succession of standard events, interspersed by the appearance of a dynamically growing oddball – that is to say, an unexpected event – of variable clock duration. Participants were asked to judge whether the oddball (to which more attention was given than to the standard events) lasted longer or shorter than the standard events. The results showed the expected subjective temporal distortions: An unexpected, oddball event draws more attention to itself than an expected, standard one. More attention correlated with subjective time dilation, less attention with time contraction. Participants overestimated the duration of the oddball events and underestimated that of the standard ones. But, at the very start of each exposure to a new stimulus, there was found to be a threshold below which distortions did not occur in the expected way. For the oddball, for instance,

> time does not appear to expand subjectively until 75-120ms after stimulus onset. This result is consistent with the view that duration overestimation is a function of the allocation of attention, because attention presumably takes some time to allocate the oddball target after it is detected. [11]

In fact, Tse observed an underestimation of duration below this onset threshold. In terms of fractal time, this threshold can be interpreted as the transition period in which new LODs are generated. The underestimation of duration during that interval may be due to the generation of Δt_{depth} as a result of new nesting. A shift in attention and the accompanying re-contextualization do not occur instantaneously. The stimulus in focus – the new content – is embedded into a context. In other words, we create Δt_{depth} (simultaneity) and time speeds up. Once the new LOD exists and is exclusively focussed on, we generate Δt_{length} (succession) only and time slows down.

There is abundant literature on the effect of attention on temporal perception. The detail and differentiated approaches of this literature are beyond the scope of this book. Here, it shall suffice to give a fractal interpretation of the relation between attention and temporal distortion in terms of contextualization and de-contextualization – the generation of Δt_{depth} and Δt_{length}, respectively.

Another powerful context which distorts our judgement of duration is emotion and empathy. Sylvie Droit-Volet and Sandrine Gil have shown how intrinsic context (an emotional state) distorts our estimation of duration and extrinsic context (other people's rhythms) changes our own temporal rhythm during entrainment [12]. Droit-Volet and Gil stress that such temporal distortions are by no means a sign that our internal clock is dysfunctional. On the contrary, they are a sign that we can adapt in a complex manner to events in our environment by distorting the speed of the passage of time or locking into other peoples' rhythms.

> (...) there is no unique, homogeneous time but instead multiple experiences of time. Our subjective temporal distortions directly reflect the way our brain and body adapt to these multiple time scales. [13]

This approach may be said to support the fractal time model, as it considers multiple time scales to which our brains and bodies adapt. Although Droit-Volet and Gil do not explicitly refer to nested scales, the fractal idea is implicit in the simultaneous multi-scale physical and mental interactions with our environment.

Droit-Volet and Gil differentiate between an attention-driven mechanism for low arousal and an emotion-driven one for high arousal. As we have seen, attention usually means focussing on a detail (one LOD) and giving less or no attention to context – so time slows down. The opposite seems to be true for emotions, as high-arousal stimuli activate the entire body by increasing heart rate and blood pressure, contracting muscles, etc. This is important when we suddenly have to flee or fight, but also, to a lesser extent, when we prepare for a positive interaction. So high arousal leads to activity on multiple nested time scales within the body.

The influence of emotions on time perception depends not only on the arousal level, but is also determined by each discrete emotion. In an additional experiment, Droit-Volet and Gil presented participants with faces with a neutral expression and faces which expressed a basic emotion – that is to say, anger, fear, happiness, sadness or disgust [14]. They then had to rate the duration of exposure of each face as long or short (this was done by priming them with two anchor durations, a short one and a long one, before the experiment). It turned out that participants rated angry and fearful faces as 'long' more often than neutral ones. This may not come as a complete surprise, as anger and fear are closely interlinked with stress, which – as we have seen – invariably leads to time dilation. Not all negative emotions, however, lead to a subjective slowing down of time. In a recent study, Droit-Volet and Meck showed that the emotional facial expression of disgust does not cause any temporal distortion, although it is both unpleasant and a high-arousal emotion, as fear and anger are [15]. So it is necessary to take account of the particular emotion and its object (to which an emotion is always linked, in contrast to a mood, which is diffuse [16]). Consideration must be given, too, to the arousal level, and to other relevant factors which may be influencing our perception in this context.

Apart from intrinsic context, such as emotion, our perception of time is also highly susceptible to extrinsic contexts, such as the speed and rhythm of other people. We tend to lock into the rhythms of others and to incorporate their time. This is not surprising, as successful interaction with our environment heavily depends on our ability to synchronize our own rhythm and speed with those of the individuals with whom we interact. The more we like someone, the more likely we are to imitate that person's rhythm and timing: for it has long been held that "imitation is the sincerest of flattery" [17]. This effect can serve as a measure of successful human interaction. In a recent study, Fabian Ramseyer and Wolfgang Tschacher have shown that the success of psychotherapy sessions correlates with the degree of synchronization between therapist and client [18]. By means of a methodological approach known as motion energy analysis, based on the processing of filmed sequences of interaction in psychotherapy, a correlation was found between nonverbal synchrony (motion and gesture imitation between therapist and client) and the quality of the therapeutic bond (as judged from the therapist's perspective). A high degree of nonverbal synchrony corresponded to a strong therapeutic bond, low synchrony to a weak one. Such entrainment of nonverbal expressions occurs involuntarily and usually goes unnoticed by the individuals involved. Droit-Volet and Gil report a study in which students involuntarily locked into the (slower) rhythms of elderly people:

> (...) the subjects had to form sentences from a list of words. In the control group, the words were neutral, while, in the experimental group, a subset of the words related to elderly traits, e.g., grey, bingo, were used. When they left the laboratory to reach the elevator, the students primed with the elderly category walked more slowly than the non-primed students. [19]

The students simulated the bodily state of the elderly people, which slowed down their own movement. It has long been known that our mirror neurons simulate extrinsic contexts, such as the facial expressions or movements of others [20]. We produce the same emotions which we perceive in others because our mirror neurons fire in almost the same way, whether we experience this emotion ourselves or observe it in someone else. An embodied participant, for instance, a human being, with a physical body nested into a physical context, can incorporate the bodily states of other people. He may, however, also fail to do so or consciously reject contextualization altogether. People who are incapable of empathy will miss out on vital communication clues and will, therefore, interact less successfully, The reason for lack of empathy and entrainment may be a pathology like autism or the desensitizing effect of prolonged exposure to stressful stimuli (which, as we saw in Chapter 5), was the case with the doctors whose mirror neurons ceased to function when they unlearned to empathize with their patient's plights [21]. Albeit to a lesser degree, we are all biased to some extent to empathize with some people more than with others.

Droit-Volet and Gil have shown that people are more likely to imitate emotions and gestures of individuals with whom they identify or whom they find appealing. There seems to be a limit to the degree to which we match our speed and rhythm to that of others and it seems that emotional contagion (and, in the wake of it, contagion of speed and rhythm) works better if the people who mimic are of the same sex as those they observe. In a separate study in which Droit-Volet and Gil were involved, Modillon et al asked participants to look at pictures of angry and neutral faces of both Chinese and Caucasian people and rate the duration of the exposure [22]. Caucasian participants overestimated the duration of the exposure of angry Caucasian faces, but not that of the Chinese faces. So there was an in-group effect which made them more likely to empathize with Caucasian faces. A high level of empathy produced an overestimation of time, which resulted from participants' ability to identify with the angry faces and thus adopt their speed and rhythms by imitation.

But our ability to empathize can also be inhibited by modifying posture or facial expressions. The muscles which we use in smiling can be activated by holding a pen between the lips. This can lead us to interpret faces as happier and cartoons as funnier than if those muscles were not activated [23]. The limiting effect of this was shown in a recent experiment, when one group of participants was able to react spontaneously and imitate the emotion on the face shown, while the others were asked to hold a pen between their lips (which biased them towards a happier interpretation of whatever they encountered) [24]. The responses of the inhibited group, which could not imitate what they saw, showed almost no difference in their reactions to emotional and neutral faces. This also prevented them from locking into the temporal rhythms associated with emotional faces. Whereas the uninhibited group was able to empathize with the faces on the pictures and locked into their temporal rhythms, the group which was inhibited from reacting spontaneously did not mimic the facial expressions and therefore experienced no temporal distortion. Droit-Volet and Gil conclude that

(...) the representation of a particular duration is highly context dependent: It depends on both intrinsic context, such as the emotional state at the onset of time processing, and extrinsic context, such as the others' activity rhythm. Our studies also suggest that these contextual variations of subjective time do not result from the incorrect functioning of the internal clock but, on the contrary, from the excellent ability of the internal clock to adapt to events in the environment. (...) Our temporal distortions directly reflect the way our brain and body adapt to (...) multiple times. [25]

The successful matching of external and internal speed and rhythm facilitates communication. It allows us to anticipate turning signals within a conversation and thus successfully interact with each other. So our ability to adapt depends on the degree to which we are able to imitate external speeds and rhythms. A high level of possible temporal responses means a high degree of internal complexity – the condition which Bruce West has equated with a state of health [26].

Emotion also has a strong influence on memory formation, in the role of both content and context. But, as Nicholas Medford et al indicate, although "there is strong evidence that memory is enhanced by emotional arousal at the time of encoding (...), the specific neural and cognitive mechanisms remain elusive." [27] Having studied the neural correlates of memory for both content and context, Medford et al conclude that they are processed in different parts of the brain. On the behavioural level, this finding is supported by he fact that

After an emotional event, while recall of the event itself is enhanced, memory for surrounding contextual information is variously found to be either enhanced or diminished, raising the possibility that memory for content and context are subserved by different neural networks. [28]

In an experiment, Medford et al asked participants to silently read a list of sentence pairs. Each pair differed by one word, which was either neutral or emotionally aversive. So, for instance, while the neutral sentence read "He stood on the balcony and watched the tide," its emotional counterpart would read "He stood on the balcony and watched the riot." The participants would later have to remember an embedded word: in this case "balcony". After reading the sentence pairs, participants were given three words, of which they had only seen one, and had to remember which they had encountered earlier, embedded in a sentence. (Of the words used to create the groups of three, half were emotionally aversive and half were neutral). It turned out that the participants had remembered emotionally aversive words better than neutral ones. But, in addition, context also influenced the memory of the neutral words which were embedded in emotional sentences immediately before or after the aversive word. Medford et al conclude that

> (...) emotionally salient verbal material is better remembered than affectively neutral, but otherwise similar, material. This effect extends to context words as well as words that are emotionally salient. [29]

Thus, aversive emotional context enhances memory, regardless of content. From a fractal perspective, emotion provides an additional context, an extra LOD, during memory formation and thereby creates Δt_{depth}. However, as we saw earlier, it is not only the emotional content or context which distorts our perception of duration. The level of arousal has at least as much, if not more, impact on our judgement of duration. This has also been shown by Alessandro Angrilli et al, who asked participants in an experiment to judge the duration of slides with emotional and neutral images. It turned out that

> for low arousal stimuli, the duration of negative slides was judged relatively shorter than the duration of positive slides. For high arousal stimuli, the duration of negative slides was judged longer than the duration of positive slides. [30]

The authors also assumed the two different motivational mechanisms – the emotional and the attentional – to be at work, both of which are controlled by the level of arousal (see Fig. 8.3).

The idea that increased attention causes an increase in perceived duration has been challenged by Eagleman, who proposes that it is actually predictability which is responsible for this phenomenon. Although Eagleman's initial finding that "the subjective duration assigned to a stimulus reflects the magnitude of the neural response to the stimulus" appears, at first sight, to support attention theory, he comes up with a plausible alternative interpretation:

> While it has been previously assumed that increases in attention are responsible for increases in duration (...), we would like to at least consider the opposite hypothesis: perhaps increases in subjective duration drive attention, allowing more opportunity for the perceptual systems to 'grab onto' a stimulus. Note that the correlation between duration and attention does not specify a causal direction, so it is logically possible that the direction goes either way. (...) In other words, unpredictability allows the attentional machinery more opportunity to fixate on the new data in order to be better positioned to predict it upon the next encounter. [31]

The first part of Eagleman's interpretation – that is to say, the idea that an increase in subjective duration drives attention – would also allow for a reversal of the causal order of focus and time dilation in meditation. Unpredictability, however, is specifically avoided in meditation. On the contrary, it is the focus on repetitive, periodic structures (e.g. humming a mantra) which induces the focussed state. Eagleman's interpretation may work well if we link high and low arousal levels to the degree of predictability and settle meditation on the low arousal side.

It is tempting to see emotions, levels of arousal and predictability as additional context which shape a content such as the estimation of a temporal interval. However, this is problematic, as there is no clear distinction between content and context if we are to believe Eagleman's suggestion. On the one hand, duration is the context for giving something our full attention but, on the other hand, as a structure shaped by the degree of unpredictability, it is also a content.

Another difficulty arises in the wake of Antonio Damasio's work: he sees emotions not just as additional context but as prerequisites for almost all cognitive functions [32]. (A well-documented example is our ability to make decisions.) The effect of emotions on cognition, as described by Damasio, is reminiscent – at least in some respects – of the effect David Bohm's implicate order (the totality of higher contexts) has on its manifestations (phenomena).

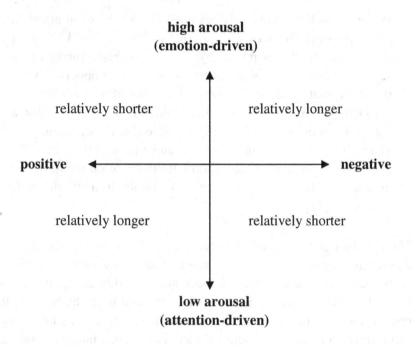

**high arousal
(emotion-driven)**

relatively shorter   relatively longer

positive ⟵——————————————⟶ **negative**

relatively longer   relatively shorter

**low arousal
(attention-driven)**

Fig. 8.3. Both arousal and affective valence affect our judgement of duration.

Both in Buddhism and in Bohm's idea of the implicate order, the reality we perceive is both determined and relativized by the distinction between content and context. If all meaning is context-dependent, all meanings are relative. Content and context correspond to Bohm's concepts of soma (the physical aspect of the world) and significance (the mental part) [33]. They can never be conceived of separately, but rather as two sides of a coin, or as two aspects of one indivisible reality. Momentary phenomena are differentiated from, and stand out against, the background of higher contexts.

In Buddhist epistemology, all meanings stand for exclusions of their opposites. Elephants, for instance, stand out from the unitary field of reality because we distinguish them from non-elephants. According to this principle of the 'exclusion of the other', all our concepts are empty, as they only make sense in the context of their opposites. Also, in the effort to identify a systemic whole, we run into difficulty in Buddhist philosophy, since there exist only momentary phenomena which unfold – instant by instant – from higher contexts. An elephant is an ephemeral concept which exists as a meaningful whole only against the background of a context [34]. A somatic configuration like an elephant

> is embedded in a contextual environment of subtle significance, which remains hidden until it is addressed. Now when we investigate the context, it also becomes a mental content that is dependent on further context, and so on indefinitely. (...) In Bohm's terminology, the universe unfolds from the totality of these higher contexts, which he calls the implicate order. [35]

Consider again my sudden awareness of the lack of context on that day when I noticed the absence of jet noise and vapour trails. I perceived a somatic configuration because I had identified the negation of that configuration. That negation was the context against which I could identify the configuration.

When we consider the example of the elephant as a somatic configuration, we may well argue: This is all very well, but if the elephant gets into a must and tramples down a whole village, it surely is a separable entity for whose identification we need no further context. But we would actually be committing a fallacy reminiscent of Woody Allen's "I am plagued by doubts. What if everything is an illusion and nothing exists? In that case, I definitely overpaid for my carpet." [36]

In Chapter 11, we shall be looking at Bohm's holistic model in more detail. We may point out here, though, that is non-reductionist and non-local, and that observer and observed, thought and reality are inextricably intertwined. The higher context, from which reality unfolds, "contains" the entities which manifest themselves to us as somatic configurations in a very specific manner: content and context form one systemic whole. Therefore, it is not surprising that it does not make sense to talk of a somatic configuration as a meaningful entity in itself.

In the fractal paradigm, a large number of simultaneous somatic configurations within the Now (high degree of Δt_{depth}) is a higher-order manifestation of the implicate order than successive percepts on one or few LODs (Δt_{length}).

So if we are willing to follow Bohm and the Buddhist argument, what is it that makes us cling to the notion of an entity which makes sense intrinsically and resists contextual changes almost unscathed. In other words: how come our bodies have discernible boundaries (although these are highly negotiable) and there is something we call 'the self'? We have dealt with the notion of boundary and self in Chapter 4 in the context of extended observers. The other side of the coin, our physical body, embedded in a physical environment, is a similar issue: Why do we cling to the notion of a separable, identifiable entity? It makes more sense to 'grasp' a whole gestalt rather than the sum of its parts because this greatly facilitates communication and interaction in everyday life. I can recognize the gestalt of my neighbour immediately and successfully exchange greetings, without getting lost in detail and context.

Human beings are open systems who regulate their internal environment by adjusting and regulating their flow equilibrium in order to maintain a stable condition [37]. This ability to maintain a systemic whole is called homeostasis (from Greek ομιοσ = similar, and ιστημι = standing still). Parameters such as body temperature, for instance, can be controlled internally or externally.

With regards to any given life system parameter, an organism may be a *conformer* or a *regulator*. On one hand, regulators try to maintain the parameter at a constant level over possibly wide ambient environmental variations. On the other hand, conformers allow the environment to determine the parameter. For instance, endothermic animals maintain a constant body temperature, while exothermic (...) animals exhibit wide body temperature variation. Examples of endothermic animals include mammals and birds, examples of exothermic animals include reptiles and some sea animals. Humans are nature's ultimate example of *regulators* because they control their parameter in a variety in climates and conditions. [38]

An example of a conformer-regulator hybrid in humans is circadian rhythms (the sleep-wake cycle is influenced by the alternating periods of light and darkness). These are part of the multi-level entrainment which feeds homeostasis – the maintenance of embedded and entrained internal and external rhythms. Homeostatic imbalance almost always leads to disease, as when the body contains too much of a substance. Such is the case with metabolic diseases like gout or diabetes. In the next chapter, we shall look at dynamical diseases, which stem from temporal homeostatic imbalance. We shall deal, too, with 'temporal misfits', including people who are incapable of making up for delays by means of regulatory anticipation – that is, by incorporating temporal context into their Now.

We have seen that temporal representation is context-dependent. A cascade of nested contexts generates a fractal structure, with each embedding level providing the next-higher context for the nested structures. Subjective duration is determined by the way percepts are internally arranged. If they are assigned the status of context, the dimension of simultaneity, Δt_{depth}, is generated and duration increases. If they are given the status of content, the dimension of succession, Δt_{length}, is created and duration decreases. Table 8.1. shows correlations between the two temporal dimensions and the various influences on time perception dealt with in this chapter.

Table 8.1. Correlations between the two temporal dimensions, content and context, attention and emotion-based arousal levels and valence, entrainment, predictability and our perception of duration.

$+ \Delta t_{depth}$ $- \Delta t_{length}$	$- \Delta t_{depth}$ $+ \Delta t_{length}$
+ context	- context
- content	+ content
time speeds up (duration decreases)	time slows down (duration increases)
distributed attention	focussed attention
high-arousal (emotion-driven mechanism) and positive valence: relatively shorter	high-arousal (emotion-driven mechanism) and negative valence: relatively longer
low-arousal (attention-driven mechanism) and negative valence: relatively shorter	low-arousal (attention-driven mechanism) and positive valence: relatively longer
- entrainment (no or low level of empathy and no incorporation of extrinsic speed or rhythm)	+ entrainment (emotional contagion: high level of empathy and incorporation of others' speed and rhythm)
predictability (no new nesting)	unpredictability (new nesting)

References

1. Illustration *Rubin's Illusion* courtesy of Manisha Pandit 2010.
2. R.R. Smith, Pop as Modernism, in: The Quietus.com 17 November 2009.
3. ibid.
4. ibid.
5. K. Yarrow, Temporal Dilation: the chronostastis illusion and spatial attention, in: A.C. Nobre and J.T. Coull (Eds), *Attention and Time*, Oxford University Press 2010.
6. ibid, p. 164.
7. P.U. Tse, Attention Underlies Subjective Temporal Expansion, in: A.C. Nobre and J.C. Coull (Eds), *Attention and Time*, Oxford University Press, 2010, p. 137.
8. ibid, p. 138.
9. The Eagleman Lab (http://neuro.bcm.edu/eagleman/time.html)
10. P.U. Tse, Attention Underlies Subjective Temporal Expansion, in: A.C. Nobre and J.C. Coull (Eds), *Attention and Time*, Oxford University Press, 2010, pp. 140ff.
11. ibid, p. 140.

12. S. Droit-Volet and S. Gil, The Time-emotion paradox (review), in: *Philosophical Transactions of the Royal Society B* (2009), Vol. 364, pp. 1943-1953.
13. ibid, p. 1943.
14. S. Droit-Volet and S. Gil, The Time-emotion paradox (review), in: *Philosophical Transactions of the Royal Society B* (2009), Vol. 364, p. 1950.
15. S. Droit-Volet and W.H. Meck, How Emotions Colour our Time Perception, in *Trends Cogn. Sci.* (2007) Vol. 11, pp. 504-513.
16. G.L. Clore and K. Gasper, Feeling is Believing: Some Affective Influences on Belief, in: N.H. Frijda, A.S.R. Manstead and S. Bem (Eds.) *Emotions and beliefs: How do emotions influence beliefs?* Cambridge University Press, 2000, pp. 10-44.
17. Charles Caleb Colton (1780-1832), *Lacon*, Vol. I, no. 183.
18. F. Ramseyer and W. Tschacher, Synchrony in Dyadic Psychotherapy Sessions, in: S. Vrobel, O.E. Rössler and T. Marks-Tarlow, *Simultaneity: Temporal Structures and Observer Perspectives*, World Scientific 2008.
19. S. Droit-Volet and S. Gil, The Time-emotion paradox (review), in: *Philosophical Transactions of the Royal Society B* (2009), Vol. 364, p. 1949.
20. G. Rizolatti, *Mirrors in the Brain – How our Minds Share Actions and Emotions*. Oxford University Press 2008.
21. Y. Cheng, C. Lin, H.L. Liu, Y. Hsu, D. Hung, J. Decety, Expertise Modulates the Perception of Pain in Others. In: *Current Biology*. Sept. 2007, pp. 1708-1713.
22. L. Modillon, P.M. Niedenthal, S. Gil and S. Droit-Volet, Imitation of in-group versus out-group members' facial expressions of anger: a test with a time perception task, in: *Soc. Neurosci.* (2007) Vol. 2, pp. 223-237.
23. Maja Storch et al: *Embodiment Die Wechselwirkung von Körper und Psyche verstehen und nutzen*. Huber, Bern 2006.
24. D. Effron, P.M. Niedenthal, S. Gil and S. Droit-Volet, Embodied Temporal Perception of Emotion. *Emotion* (2006) Vol. 6, pp. 1-9.
25. S. Droit-Volet and S. Gil, The Time-emotion paradox (review), in: *Philosophical Transactions of the Royal Society B* (2009), Vol. 364, p. 1950.
26. B.J. West, *Where Medicine Went Wrong – Rediscovering the Path to Complexity*, World Scientific, Singapore 2006.
27. N. Medford et al, Emotional Memory: Separating Content and Context, in: *Psychiatric Research: Neuroimaging* (2005) Vol. 138, p. 247.
28. ibid, p. 248.
29. ibid, p. 253.
30. A. Angrilli, P. Cherubini, A. Pavese and S. Manfredini, The Influence of Affective Factors on Time Perception, in: *Perception & Psychophysics* (1997), 59 (6), pp. 972-982.
31. D.M. Eagleman, Duration Illusions and Predictability, in: A.C. Nobre and J.C. Coull (Eds), *Attention and Time*, Oxford University Press, 2010, p. 157.
32. A. Damasio, *Descartes' Error: Emotion, Reason, and the Human Brain*. Penguin, London 1994.

33. D. Bohm, *Wholeness and the Implicate Order*, Routledge, London 1980.
34. T. Agócs, The Mystery of Meaning, paper prepared for *Science and Religion: Global Perspectives*, June 4-8 2005, Metanexus Institute, Philadelphia.
35. ibid, p. 13.
36. W. Allen, *Without Feathers*, Sphere Books, London 1986, p. 6.
37. *Wikipedia*: Homeostasis, 29 May 2010.
38. ibid.

Chapter 9

Modifying Duration I:
Nesting and De-Nesting

In the last chapter, we saw how emotions and temporal distortions associated with them can be contagious, provided our empathic abilities are not compromised. But, if a current trend continues, rhythms may no longer rub off. According to a study conducted by Sara Konrath et al, the empathic skills of American college kids have dropped by 40 percent over a period of 20 or 30 years ago [1, 2]. The biggest fall occurred between 2000 and 2009, suggesting a trend. The reason for this sharp decline in a vital ability is not clear. However, the usual suspects have been rounded up, such as the shifting of physical meetings to virtual encounters and exposure to violent media, which desensitizes people to the pain of others. As with doctors who have become benumbed by prolonged exposure to patients' pain or people who are capable of torturing or killing others, those students' mirror neurons seem to no longer perform the task of simulating human behaviour in their environment. This is a serious development, as empathy is a key skill which allows us to communicate successfully – that is to say, to avoid misunderstandings by reading facial expressions correctly or by anticipating coming turning signals.

A lack of empathy severely compromises our ability to lock into other people's rhythms and to profit from them. A healthy person's gait, for instance, can entrain someone who is motorically limited – say, a sufferer from from Parkinson's disease – and help that individual to walk more smoothly.

In this chapter, we shall look at how nesting and de-nesting can help people to re-attain their ability to assimilate or reject context – that is to say, to lock into or out of environmental rhythms. Then, we shall

179

investigate how the interaction between internal and external rhythms can lead to a healthy state or to what we shall define as a dynamical disease. To get an idea of what we actually know about this interaction, let us look at the related fields of chronobiology and chronopsychology.

> *Chronobiology* (...) examines physiological events in living organisms that occur on a periodical basis (so-called *biological rhythms*) and investigates how they are influenced by (and adapt to) external rhythms. *Chronopsychology*, on the other hand, is concerned with psychological rhythms and their influence on cognition and behaviour. [3]

Internal cycles are linked both to other endogenous rhythms and exogenous *zeitgeber* – that is to say, external synchronizing stimuli. Hormone production, for example, is influenced by an individual's sleep-wake cycle. A disruption of that rhythm has a negative impact on health, when, for instance, in children, through lack of sleep, the human growth hormone is not sufficiently activated [4]. But there are also internal rhythms – often referred to as 'inner clocks' – which seem to run independently of any external synchronizing stimulus. So we appear to be – at least to some degree – independent of extrinsic factors and, therefore, a hybrid of controller and regulator. However, it is difficult to say how those internal clocks would behave, say, after a decade of selected sensory deprivation. Possibly, they would just have to be wound up less frequently. And as we are physically and mentally embedded in our temporal and spatial environment, it would be surprising if, within us, there existed some monadic time setter totally independent of external synchronization. Although internal clocks may be ticking away undisturbed, they are still linked – albeit indirectly – not only to neuronal oscillators and metabolic rhythms, but also to environmental ones. So we should not disregard the possibility that although some cycles may need winding up only very rarely, possibly only once in a lifetime, they are never isolated from their environment. For instance, we cannot escape gravity or the Earth's rotation around the sun.

The rhythm which exerts the strongest impact on living organisms is the circadian one, which is a cycle of approximately 24 hours. In isolation experiments, it has been shown that the average circadian rhythm is closer to 25 hours, implying that it is based on an endogenous *zeitgeber* – an inner clock. This conclusion is also supported by the observation that people who live between the Arctic Circle and the North Pole sleep once a day, just like those in, say, Central Europe. However, sleeping patterns of people who live north of the Arctic Circle vary more over the year than those of people who live close to the equator [5]. Kirsten Brukamp gives an overview of processes which are directly influenced by circadian rhythms:

> Circadian rhythms determine functions such as sleep-wake cycles, periodic hormone levels, thermoregulation and appetite cycles (...). In particular, circadian clocks control the biological functions of metabolism, such as food intake, biochemical reactions of metabolism, detoxification and excretion. An example is the steering of the waste disposal systems in liver, gut and kidneys. [6]

As a further instance, we can point to detoxification, which depends strongly on circadian rhythm. Evidence of this is that some drugs reveal an astonishing range in effect (depending on the timing with which they are administered), from lethal to almost non-toxic [7].

The internal clock which keeps our circadian rhythm going is synchronized by the suprachiasmatic nucleus (SCN), a cluster of approximately 20,000 neurons situated in the hypothalamus [8]. This master pacemaker synchronizes a number of peripheral oscillators which regulate periodic gene activity. Brukamp points out the hierarchical structure of pacemaker and peripheral clocks:

> The peripheral cells receive the periodic information necessary for their synchronization from the SCN and bodily functions, such as food intake. The clock cells are thus organized in a hierarchy with the SCN at the top, although their molecular mechanisms are fairly uniform. [9]

Such peripheral oscillators have a wide-ranging impact, as they may continue to oscillate during mitosis and are passed on to daughter cells.

Disrupted circadian rhythms can be reset, for instance, by means of light therapy or by administering melatonin. Normally, our circadian rhythm is synchronized with environmental *zeitgeber* such as day and night – that is to say, the alternation of light and darkness. However, artificial *zeitgeber* can take their place, if only to a certain degree, as isolation experiments have shown: there is a 'range of entrainment' beyond which synchronization no longer occurs. Referring to Wever et al's work [10], Jürgen Zulley summarizes:

> It was found that the simple alternation of darkness with normal indoor lighting (about 400 lux) is not sufficient to entrain the normal daily rhythm. Entrainment was very much greater if the strength of the light was increased to that of broad daylight (more than 2000 lux). It therefore follows that daylight must be regarded as the most effective *zeitgeber* (...). Other important *zeitgeber* are the regular timing of the meals as well as motor activity and social contacts. [11]

Apart from circadian cycles, there are other rhythms which influence living organisms at higher (ultradian) or lower frequencies (infradian). Ultradian rhythms are shorter than 24 hours and occur on several nested timescales: for instance, the 90-minute REM cycle, the 4-hour nasal cycle and the 3-hour growth hormone cycle. Infradian rhythms are longer than a day and also include nested intervals such as the female menstrual cycle and seasonal changes in temperature and humidity [12].

It looks as if sleep-wake patterns are also connected to ultradian cycles, such as the two daily maxima of orthostatic dysregulation – that is to say, the state of less variability in the accommodation of changes in blood pressure – which occur at three in the morning and around noon. The afternoon nap is beneficial because, according to Zulley reduced vigilance, an increased rate of error performance and a lowering of body temperature – all of which are independent of food intake – indicate that the organism is experiencing rearrangements during the day similar to those which characterize the second half of the night. This suggests that a resting phase belongs to this time. The organism is therefore not strictly adjusted to constant activity with only a single recovery phase within the circadian rhythm, but shows a change during this active phase towards a period of relaxation. [13]

Zulley also found that there are even shorter periods – intervals of approximately four hours – when we feel more need for sleep [14]. In children, who need more sleep than adults, these four-hour alternations are superimposed on the circadian fluctuations. This superposition shows itself less and less as the children grow up. In advanced old age, however, a four-hour periodicity in sleeping and waking indicates a pathological condition.

Among the external *zeitgeber* acting on our bodies and psyche, social rhythms must also be mentioned. The interdisciplinary area of research called 'chronoscience' studies the general impact of rhythms. In particular, it looks at how rhythms influence our well-being, which is often disrupted when we have to lock into an external rhythm which runs against our internal ones. Of course, we can adapt, say, to shift work or getting up very early in the morning even though this may be incompatible with your chronotype. People tend to be either night owls (staying up and getting up late) or larks (going to bed and rising early). For both categories, this can lead to difficulties when they have to adapt to a timetable which does not match their chronotype. Teenagers, for instance, tend to be night owls rather than larks [15]. Yet many are forced to start lessons early in the morning, often before sunrise. Their early-morning lethargy is therefore understandable and could be remedied by shifting their timetable. However, this would be bad news for teachers over 50, as people in this age group tend to be larks. It should be mentioned, too, that the change in rhythm demanded by shift work has been shown to take its toll on human health [16].

Henri Lefebre, the inventor of 'rhythmanalysis', recognized the coexistence of our social and biological rhythms and stressed that it is the human body which is the point of contact between those rhythms [17]. He also saw that the relationship between internal and external rhythms is increasingly biased towards environmental influences: Our biological rhythms, which determine when we eat and sleep, are more and more conditioned by our work and social life. For the rhythmanalyst, the body is not the object of study, but is rather a tool to find the linkage between entrained or nested rhythms. He uses the body as a metronome – a temporal yardstick which can translate between biological and social rhythms and between inside and outside:

> Acquired rhythms are simultaneously internal and social. In one day in the modern world, everybody does more or less the same thing at more or less the same times, but each person is really alone in doing it. [18]

Lefebre describes the beauty and harmony of the human body, calling it a bundle, a bouquet, a garland of rhythms, and ponders the origin of the human ability to adapt and balance those multiple simultaneous rhythms:

> The living – polyrhythmic – body is composed of diverse rhythms, each 'part', each organ or function having its own, in a perpetual interaction, in a doubtlessly 'metastable' equilibrium, always compromised, though usually recovered, except in cases of disruption. How? By a simple mechanism? By homeostasis, as in cybernetics? Or more subtly, through a hierarchical arrangement of centres, with one higher centre giving order to relational activity? This is one of our questions. But the surroundings of the body, the social just as much as the cosmic body, are equally bundles of rhythms (...) [19]

We find simultaneity not only between the rhythms within a human being and environmental rhythms, but also within one systemic whole, such as a tree. Lefebre points out that, for instance, a cherry tree has several rhythms, which brings forth leaves, flowers, fruit and seeds. Both flowers and leaves will appear in the spring, but the leaves will outlast the flowers until the autumn. The leaf-carrying interval thus embeds the flowering interval – a temporal fractal. Once we don fractal spectacles, we find ourselves able to perceive a symphony of simultaneity. Although Lefebre did not mention fractals, he describes nested rhythms exactly as a temporal fractal:

> From these first glimpses, the outcome is that the living body can and must consider itself as an interaction of organs situated inside it, where each organ has its own rhythm but is subjected to a spatio-temporal whole [globalité]. Furthermore, this human body is the site and place of interaction between the biological, the physiological (natural) and the social (often called the cultural), where each of these levels, each of these dimensions, has its own specificity, therefore, its space-time: its rhythm. Whence the inevitable shocks (stresses), disruptions and disturbances in this ensemble whose stability is absolutely never guaranteed. Whence the importance of scales, proportions and rhythms. [20]

Rhythmanalysis sees the body as the pivotal point where internal and external rhythms meet. We are, however, left in the dark as to exactly how body and environment meet and how the boundary between them is maintained. Otto van Nieuwenhuijze has come up with a suggestion: Our body is an equation – it equates the balance between internal and external coherence [21]. To describe van Nieuwenhuize's cosmology would go beyond the scope of this book [22]. Here, it shall suffice briefly to outline his Equation of Health, which describes the balance between extramission from within the body and intromission via impressions from our context. Any endo-exo interaction can destabilize or re-stabilize a system – in this case, the human body. An example of this is the pro/anti-inflammatory mechanism for redefining internal membranes of the system – that is to say, a localized protective response of the vascularized tissue to, for instance, wall off infected tissue.

Van Nieuwenhuijze stresses that our body expresses an equation of the internal balance and that between the part and the whole: the system (body) as part of its context. Every body cell computes the coherence of the system and the system in its context – our cells are operators. The crux of this Equation of Health is that the body itself is the only existing expression of this equation – a form of mathematical equation which equates the part and the whole.

The Equation of Health is a fractal: The embryological unfolding is a step-by-step expression of a balance. This balance is the equation which describes how the zygote – that is to say, the first cell – multiplies by means of cell division while always maintaining its inner and outer balance. Once this balance is disturbed or lost, the human being in question falls ill or dies. Van Nieuwenhuijze describes the series of cellular divisions from the zygote to the integral body as the *body diffractal*. Every cell can always be traced back through the bifurcations to the unfolding zygote.

> From the unfoldment of the zygote, the system membrane is expanded to form the skin of the body. In mathematical terms this can be regarded as the wave envelope for a wave group. The sensory capacities of the body are expressions of the initial sensitivity of the membrane of the zygote. In the same way as the zygote unfolds, the contact with the context feeds back to the point of origination. [23]

Health can only be maintained if the body is nested into a healthy social and environmental context. When balance is sought, all aspects of our physical and mental involvement need to be considered simultaneously. And, conversely, balanced sovereign individuals are required for successful communication to take place. Van Nieuwenhuijze stresses that inter-human communication is based on intra-human (cell) communication [24]. Our degree of involvement determines whether we are healthy and live in peaceful relationships with others. This involvement occurs on several levels of description simultaneously: our mental, spiritual and physical dimensions all interact with our environment.

When we communicate with others, we have to cross the boundary which defines our physical and mental extensions. In order to do this, according to van Nieuwenhuijze, we have to take account of a number of boundaries within and outside the body, including our skin, the interspace between people and the external space which is beyond our reach. With each boundary crossed, we lose a degree of freedom and, therefore, control. By the time we are interacting with another human being, we have lost our degrees of freedom, which means we have no knowledge of, and cannot control, the other person. It is only when we recognize the total sovereignty of the other that peaceful communication is possible. It follows that autarchy can only grow if sovereignty is respected among all beings. According to van Nieuwenhuijze, the formula for peace is the realization that the individual can use the freedom of choice within him, but that no individual has freedom of choice regarding others.

If the delicate balance between a person and his social environment is disturbed in one or several of the dimensions of interaction, that individual will fall ill or die and, in doing so, will also destabilize his environment. It is, furthermore, when we fail to realize that we have no control over others that we generate dis-ease and conflict.

Endo-exo mismatches often manifest themselves as pathological conditions. A wide range of disorders and diseases such as Morbus Parkinson, epilepsy, schizophrenia, depression and mania have now been categorized as dynamical diseases – a term coined by Glass and Mackey [25]. A dynamical disease is present when the body's own healthy dynamics is re-organized by temporal patterns that induce pathological rhythms. Such diseases cannot be diagnosed by focussing on a single rhythm. Instead, it is necessary to study all or many parallel rhythms of the body to detect dynamical changes that indicate pathological behaviour. Gerold Baier and Thomas Hermann suggest the use of sonification – that is to say, the auditory inspection of data in real time or in retrospect – to reveal the temporal inter-dependence of such multi-scale physical events [26].

The rhythms can be of very different temporal scale, ranging from milliseconds for neural firing to minutes for intracellular messenger signalling to hours for episodic hormone release to years in growth and aging. In addition, they tend to have complex relationships among themselves (somewhere between independent and fully synchronized) as, for example, in the relationship between breathing and the heartbeat. [27]

Sonification allows us to synthesize sounds from any data string – it is not limited to processes which generate pressure waves in their surrounding medium. An example of such processes which do not generate sound but, in this case, electric potentials, is neural activity in the brain. Event-based sonification has been used to monitor, for instance, the abnormal rhythms in epilepsy. It is a particularly useful method because it can differentiate between normal background activity and pathological rhythms. Baier and Hermann have defined an event as local maxima in two time series recorded in parallel. In addition, the amplitude of an oscillation is represented by volume and the temporal differences between the maximum on one channel to the preceding maximum on the other channel are represented by the number of harmonics. So maxima in the two channels which are divided by large intervals would produce a dull sound, whereas a bright, sharp sound would represent a short separating interval or simultaneity.

Background activity is characterized by its ongoing, irregularly occurring sound events. Both the timing and the volume distribution resemble a stochastic process. No clear rhythmic component is perceptible. The seizure activity, by contrast, is recognizable by way of its increased intensities of the beats and its pronounced regularity. A punctuated rhythm consisting of a repeating series of one fast and one slow interbeat interval is salient. The transition from and to background activity marks the beginning and the end of the attack. As is typical for absence seizures, the changes occur in both recorded channels. Thus, by virtue of the sonification, the epileptic seizure becomes an audibly perceptible unit or gestalt. [28]

Baier and Hermann suggest that multi-level sonification which makes use of more than two electrode recordings (in their case, the original recording of the electric potentials being registered by 19 electrodes), larger gestalts in the form of waves from the frontal to the occipital regions of the scalp could be made audible in soundscapes as virtual auditory environments produced by multi-speaker outputs. This would allow a very intuitive and direct way of differentiating pathological from normal rhythms. And as Baier and Hermann have generated sonifications of abnormal rhythms which were recorded when no clinical seizure occurred, it would prove to be a very differentiated non-invasive method of diagnostics. One reason why sonification is such a useful instrument of mediation is that human beings are very good at identifying auditory gestalts in real time. We are highly skillful at perceiving complex temporal correlation patterns and at generating auditory gestalts – that is to say, higher-level information – from successive and simultaneous events:

> The analysis of complex auto-correlations (correlations between successive events in a single sound source) is performed almost without effort in speech perception and the analysis of cross-correlations (correlations between events in distinct sound sources) is continuously at work in the perception of music with multiple voices, be it homophonic or polyphonic. In particular, the latter feature may prove valuable when applied to multivariate physiologic data where there is complex interplay between events and where there are interactions on different time scales. [29]

Sonification is an ideal mediator of bodily rhythms nested on various time scales, as our auditory perspective is naturally shaped by a fractal temporal structure: the signals fall onto – so to speak – fertile ears, because we are used to perceiving a fractal ensemble of sounds, of which we are capable of isolating relevant ones and assigning others the status of background noise. As Baier and Hermann have pointed out, we have come a long way from the first mediator of sounds from within the body: the stethoscope, which Rene Laennec invented for 'mediate auscultation' – that is to say, listening to sounds spontaneously produced by the body – to sonification. Mediate auscultation was used to interpret the sounds of respiration because the lungs resonate the body's own muscular activity [30]. So if someone's lungs were damaged or filled with water, his breathing would also sound different. However, this method did not focus on rhythms – it basically disregarded temporal variations of sound qualities. Event-based sonification, by contrast, allows us to "better appreciate the full spatio-temporal complexity of both healthy and pathologic physiology". [31]

The nested auditory temporal perspective described by Baier and Hermann is an example of how we can exploit our natural tendency to integrate stimuli on various time scales for diagnostic purposes. Understanding our own nested perspective helps us to describe, understand and possibly cure pathological conditions in which people are faced with a distorted temporal perspective. It was Laurent Nottale and Pierre Timar who first drew my attention to the fact that there is a connection between our fractal perspective – that is to say, scale relativity – and the spatio-temporal distortions of Parkinson's disease [32]. Morbus Parkinson, a progressive neurodegenerative disorder of the central nervous system, manifests itself as severe limitations of movement. In the appendices to his book *Awakenings*, which describes how L-DOPA and entrainment can "awake" people who suffer from Parkinson or post-encephalitic patients, Oliver Sacks refers to Parkinson as a dynamical disease [33]. He had observed that sufferers are easily catapulted from profound akineasia – that is to say, their motor activities come to a halt and they suddenly freeze – into normality and vice versa. Miniscule amounts of change in the L-DOPA doses could trigger extreme changes in behaviour. This sensitive dependence on initial conditions, which was first described by Ed Lorenz in 1963, and has since been known as the "butterfly effect', was the signature of a chaotic system. When Sacks learned about chaos theory with its strange attractors – that is to say, fractal trajectories in phase space which describe the dynamics of a system – he realized that this was the temporal structure he had been looking for in order to describe the underlying order of Parkinsonism (Parkinsonism is not idiopathic, but shares most manifestations with Parkinson's disease):

> It is as if Parkinson itself can be visualized as a sort of surface, bipolar, like a figure-of-eight. The transformations we see with L-DOPA are already inherent in Parkinsonism itself, as if L-DOPA can serve to release a tendency already built into the topological shape of Parkinsonism; or, more likely, to change the shape of the (Parkinsonian) attractor, making it sharper or steeper, with higher peaks and deeper valleys. The patient with Parkinsonism is, so to speak, enthralled on this surface, which is a dynamical surface, an orbiting surface in time. Every so often, this orbiting will take him, tantalisingly briefly, through a few seconds of 'normality', or into the mirror-state of hyperkineasia, only to return him to the immense 'gravitational' attraction of that powerful attractor which (in dynamical terms) is the 'cause' of Parkinsonism. [34]

Parkinsonian patients often perform the right movement, but on the wrong scale: They may start writing a note in normal letter-size and then change their letters to enormous or microscopic size. Likewise, they may start clapping at regular speed and then fall into festination – that is to say, clapping their hands faster and faster, with smaller and smaller gestures. The speech of Parkinsonian patients, or their gait, may also slow down or speed up, while they are usually blissfully unaware of their spatio-temporal distortions. Oliver Sacks describes how such a patient experienced and became aware of his spatial distortions: Seymour L. had been walking down the corridor normally when he suddenly accelerated, festinant, almost fell and then recovered and proceeded at normal gait. He complained to the nurse that there was suddenly a big hole in the floor: why had she not done anything about it? When assured that the floor hadn't changed, he became furious, telling her that someone must have been excavating, as there was a great hole there. He described how he was walking along, when suddenly the ground fell from his feet at a crazy angle, so that he was thrown into a run. At this point, Sacks, who had overheard he conversation, intervened and suggested the three of them walk back together to find our about the excavation. Seymour agreed, so the other two took him between them and they walked back without incident.

> The absence of incident left Seymour confounded. 'I'll be damned,' he said. 'You're perfectly right. The passage is quite level. But' – he turned to me, and spoke with an emphasis and a conviction I have never forgotten – 'I could have *sworn* it suddenly dipped, just as I said. It was *because* it dipped that I was forced into a run. You'd do the same if you felt the ground falling away, in a steep slope, from under your feet! I ran as anyone would run, with such a feeling. What you call "festination" is no more than a normal reaction to an abnormal perception. [35]

Apart from spatial distortions, Parkinsonian patients can also experience temporal dilation and contraction. Festination – a shortening of stride and quickening of gait – occurs when their spatio-temporal co-ordination system contracts. When the units of their reference frame shrink to next to nothing, akineasia occurs.

The idea is reminiscent of Zeno's paradox of motion: Achilles has to reach an infinite number of points which the tortoise has already covered, therefore he can never overtake it. And, if their steps become smaller and smaller – as in festination – the motion must necessarily grind to a halt. The Parkinsonian patient is actually caught in a series of steps of ever-diminishing size, as if his event horizon is contracting. Before he can get halfway to his goal, he has to cover a quarter of the way there, and before that, he has to cover an eighth, a sixteenth, and so on ad infinitum: a walk through fractal space-time.

Nottale and Timar's observation that the distortions in Parkinsonian patients and their motion responses can be described by the principle of scale relativity was triggered, in particular, by the fact that such patients perform the right movements on the wrong scale [36]. One of the many aspects of Nottale's Theory of Scale Relativity is that, in terms of an observer perspective, the scales of length and time, usually attributed to the observed object as being intrinsic to it, have actually no existence by themselves, since only the ratio between an external scale and an internal scale, which serves as a unit, is meaningful. [37]

Nottale's fractal space-time is continuous but nondifferentiable – in other words, it has a fractal structure. It is beyond the scope of this book to describe Nottale's theory in detail. (The interested reader is referred to Appendix D). Here, it is of particular interest that he and Timar have suggested using the scale-relativistic approach and principle to help to understand Parkinson's disease at a fundamental level. Looking at festination, they propose:

> We think that the loss of the internal device that synchronizes us to external space-time would be expressed by chaotic-like perception of dilation and contraction. (...) The behaviour of PD patients indeed seems to definitely come under the principle of scale. [38]

In a nutshell, they consider the geodesics of a fractal space-time as the possible routes on which movement may occur.

Both Parkinsonian patients and healthy people are unable to distinguish a deformation of the units of the outside world from a deformation of those of their internal structures. (This observation by Nottale is actually an extension of Boscovich invariance: see Chapter 4) Nottale and Timar stress that "the difference between the unaffected and the affected subject is therefore only a question of synchronization (or of desynchronization) of internal rulers and clocks relative to external ones." [39]

This synchronization can be triggered from the outside or, for instance, from the involuntary initiation of music which only the affected individual can hear. But because many Parkinsonian patients cannot initiate the stepping response, the entraining rhythms must be provided from the outside. A human being with a healthy stride serves as a good pacemaker. Mechanical pacemakers often have a negative effect, as they operate on one LOD only and thus narrow the walker's range of responses. In this connection, Sacks concludes that it is the lack of a healthy 'will' which distorts the topology of space-time.

> The Parkinsonian engrossed in his Parkinsonism can no more judge the 'abnormality' of his own state than could a self-conscious rod, accelerated towards the velocity of light, which had itself undergone a Lorentz contraction. It is for reasons of this sort that it is insufficient to speak of Parkinsonism as a simple 'dyskinesia' (motor disorder). One must speak of it as a systematic disorder of space-time parameters, a systematic warping of coordinate-systems; indeed one must go further, and say that this misjudgement or warping is secondary to a systematic disorder of 'will' or force, which has the effect of warping Parkinsonian 'space' and making it a dynamic, field or relativistic disorder. [40]

For an external 'will' to trigger a reaction in a Parkinsonian patient, an *embedding* rhythm must be present. However, not any rhythm will do: if it is of the wrong speed or too overpowering, it can cause a pathological response which manifests itself in ungraceful and jerky movements.

I have been describing the spatio-temporal distortions of Parkinsonian patients in detail in connection with Nottale's interpretation, which is based on his Theory of Scale Relativity. The reason for this is that it leads to the intriguing idea that healing could just be a matter of catapulting the patient onto the right spatio-temporal scale by means of simple entrainment measures and thus giving him a larger repertoire of possible responses.

However, things are not quite as simple. Often, exposing someone to an external pacemaker will have no effects or even disastrous ones – that is to say, if the pacemaker sends out a one-dimensional signal. Fractal pacemakers – ideally in the shape of human activity – trigger a range of responses on nested LODs and are therefore more suitable for restoring internal complexity and balance.

Bruce West emphasises that walking is a complex activity which involves interaction between multiple sensor systems, each operating at their own frequency, thus producing a fractal time series [41]. This time series is generated from data representing stride rate variability (SRV) – a measure which decreases with aging and as a result of some neurodegenerative diseases. West stresses that SRV is a good measure of a healthy gait and that a loss of complexity is a sign of the onset of a disorder or a disease. But not only aging and neurodegenerative diseases influence gait: West points out the additional impact of stress:

> The apparently regular stride experienced during normal walking has random fluctuations with long-term memory that persists over hundreds of steps. As a person changes their average rate of walking, either walking faster or walking slower, the scaling exponent of the fractal time series of the intervals between the strides changes with increasing stress. [42]

He points out that it is important to keep in mind that variability relates to fluctuations in the stride interval, not the interval itself. So what are remembered for a hundred strides are the past fluctuations, which are communicated through multi-level feedback control to the present fluctuations. Fractality, which is based on long-term correlations, is lost if an external pacemaker overrides the rhythm of natural gait:

> When the pace is constrained by the metronome, the stochastic properties of the stride interval time series change significantly. (...) By keying the system to respond at a set rate, the level of randomness in the allometric control of gait increases rather than decreases. The response to stress is therefore to induce loss of long-term memory. [43]

Whereas the time series of a freely walking healthy individual is multi-fractal, the impact of stress shifts the peak of the multifractal spectrum and changes its width.

West suggests that "the metronome is disrupting the usual neurological chain of command. In particular, the metronome is overriding the source of the normal long-time memory in the SRV time series" [44] and he concludes that "the traditional therapy for convalescence using machines such as stationary bicycles and tread mills are probably not the best for a rapid and complete recovery." [45]

A walker with a healthy gait, on the other hand, is a perfect time setter and can trigger a stepping response in people who are unable to set their own time spontaneously. In addition, external visual, auditory or proprioceptive cues can trigger normal movement [46]. There is plenty of evidence that, for instance, music therapy can be highly effective in re-enabling people to walk with a larger stride and at their preferred speed without the risk of falling [47, 48]. I suggest this may be so because music is not one-dimensional but has a fractal dimension – it creates Δt_{depth}.

It seems to be a matter of finding the right scale – that is to say, the right level of complexity, between rigid order and randomness, between depression and manic episodes, boredom and surprise, between time dilation and contraction. In fact, in the case of Parkinsonian patients, those corrective distortions are healthy reactions – the sufferers have a variety of reactions at their disposal to the environmental changes they perceive, festination being one of them. The reason why their internal complexity does not suffice to navigate the world is that their space-time geodesics, to use Nottale's model, do not match those their environment selects or is subjected to.

Is there a universal scale which generates something like wellness in people? There are hints that a fractal dimension around 1.3 correlates with healthy behaviour and environment. We have seen in Chapter 6 that humans show an aesthetic preference for spatial structures which display a fractal dimension of approximately 1.3., for instance, in the appreciation of paintings. We have also speculated on the restorative effect of pink noise – a temporal fractal structure present in a wide range of music. Furthermore, spatial fractals in the low to intermediate range may reduce stress. West reports that a healthy heart correlates with a heart rate variability (HRV) time series that has a fractal dimension between 1.1 and 1.3. [49]. The dynamics of a healthy adult's gait also displays a fractal dimension of around 1.3. [50]. By contrast, pathological behaviour occurs beyond these values: congestive heart failure, for instance, is characterized by a fractal dimension of 1. And the time series of atrial fibrillation has a fractal dimension of 1.5.

However, the fractal dimensions of HRV and SRV are statistical fractals based on the description of fluctuations, whereas those measuring architectural features are based on the actual extensions of the construction, not the variations in the extensions. In addition, some of these structures are spatial, others temporal. Yet other presentations of a system's dynamics take one parameter – say the physical extension or the direction of walking – and plot it with a time delay in a phase space portrait. The structure which is formed in this manner is called an attractor – its trajectory describes all possible states of the system. One type of attractor, the so-called strange attractor, has a fractal and, often, a scaling structure. It describes deterministic chaos in a system's behaviour, which means that this system is highly sensitive to initial conditions. Such attractors may be used, for instance, to differentiate between healthy and pathological gait, as in Parkinson's disease. Max Kurz et al have shown a correlation between gait and attractor divergence.

> We evaluated the amount of attractor divergence in the young, aged and individuals with Parkinson's disease as an initial step towards developing a biomechanical metric that could be used to assess walking balance. Our results illustrated that the aged and Parkinsonian gait have a greater amount of divergence. [51]

Based on related simulations, they infer that quantifying the degree of variability in the walking pattern may provide a clinical means of determining Parkinsonian patients' probability of falling.

Koyama et al have also shown in a study based on a finger tapping test that the attractors – that is to say, the graphical representation of the underlying dynamics in phase space – of Parkinsonian and healthy individuals differ significantly. In such a test, participants are asked to hit a metal cup of about 2 cm in diameter as fast as possible at constant intervals for 15 seconds. It is well-known that, as a result of their motoric limitations, Parkinsonian patients cannot perform this task well. Koyama et al's study showed that the trajectory parallel measures of healthy subjects differed from those of Parkinsonian patients by one order of magnitude. They concluded

> that quantitative values for evaluating the symptoms of Parkinson's disease can be readily determined by analyzing the time series from a finger tapping test via the trajectory parallel measure method. [52]

Many dynamical diseases display multistability – that is to say, the ability to generate a range of rhythmic activity. That rhythmic activity will show itself in fairly stable behaviour. Different behaviours can be represented by different attractors, whose trajectories will show the limitations of the underlying system's scope of change. Suppressing pathological attractors will reduce an individual's degree of internal complexity – a reduction which is desirable in such cases.

> Knowledge of the dynamic mechanisms supporting multistability opens new horizons for medical applications. It has been intensively targeted in a search for new treatments of the medical conditions which are caused by malfunction of the dynamics of the CNS. Sudden infant death, epilepsy and Parkinson's disease are examples of such conditions. Recent progress in the modern technology of computer-brain interface based on real time systems allows us to utilize complex feedback stimulation algorithms which suppress a pathological regime co-existing with the normal one. [53]

Eliminating undesirable co-existing attractors and introducing or re-introducing desirable ones reduces or extends people's degrees of freedom – it defines their dimensionality. From what we have considered thus far, it is conceivable that what is actually happening when a multi-scale external stimulus like music allows someone in some way to internalize a rhythm is that the individual is made aware of a different scale and thus increases his internal temporal complexity. But then, this complexity is already at his disposal when he is subjected to the geometry of a dilated or contracted space-time which fits that of his perceived environment: he follows a kind of forced scale-hopping exercise, down the nested rabbit hole. As if an invisible attractor – the topological structure of a fractal space-time – pulled him onto deeper and deeper layers of the attractor, limiting his movement to increasingly shorter extensions. Therefore, the external pacemaker – for instance, an accompanying human with a healthy gait – merely catapults him onto that other person's scale, onto their counterpart of a healthy attractor.

But also if co-existing attractors are lacking, a patient's internal rhythm could be re-calibrated by means of entrainment by catapulting him onto another scale so as to increase his range of internal responses. This would be an elegant method of cure or, at least, it would provide temporary relief. Often, it makes sense to combine entrainment with an embedding (nesting) of that person into a social context. Following the 2010 earthquake in Haiti, Bernd Ruf and his team set up a safe space for displaced, traumatized children and implemented emergency education measures for them [54]. UNESCO stresses that such education "should not be seen as a relief exercise, like handing out cooking pots and blankets, but as a vital dimension of national reconstruction" [55]. The idea here is to break the cycle of violence and poverty which results from the lack of education that can affect an entire generation after a natural disaster or a civil war. Ruf and his team got into the country early enough to take measures to balance and stabilize traumatized children during the 8-week window between the traumatic event and the recommended onset of psychotherapy. Both nesting and entrainment were achieved by re-setting the children's internal rhythms on various LODs [56].

First of all, these children had to re-draw their body maps, as these were severely disjointed as a result of their traumatic ordeal. When they were first asked to draw a picture of themselves, many children would produce portraits without hands and feet, as they could actually no longer feel them. After a week of rhythmic exercises, which would consist of clapping, chanting, dancing in a group and movement therapy, such as the Feldenkrais method, the children's body maps had been successfully re-integrated. Their self-portraits showed their entire bodies, complete with hands and feet.

At the same time, longer external rhythms were set up to entrain the children's internal rhythms. These would include strict mealtimes, bedtimes and waking times. The meals themselves were also ritualized. Almost military-like measures had to be implemented to keep the children from fighting over the food or wolfing it down. When served, they would hold their plate in one hand while the other arm was at an angle behind their back. Eating was done by means of slow spooning in small groups. Breakfast (a glass of water) was followed by ritual chanting and movement, then work sessions and creative exercises filled the rest of the day. In the afternoon, a cooked meal was served. Ruf and his team stressed the importance of re-establishing a strict rhythm during the first eight weeks after the traumatic event. After this window, it would have been extremely hard, if not impossible, for the children to have re-drawn their body maps [57].

If internal rhythms are upset after a traumatic experience such as this, both psyche and body become disjointed. Linking inside and outside is the first emergency measure to be taken when internal rhythms have broken down. However, this adapting is only the first step. Ultimately, internal complexity needs to be restored not only in the individual but in society at large. For, as Sacks has observed, complex social rituals and behaviour are emergent phenomena held together by rhythms.

Neuroscientists sometimes speak of the binding problem, the process by which different perceptions or aspects of perception are bound together and unified. What enables us, for example, to bind together the sight, sound, smell, and emotions aroused by the sight of a jaguar? Such binding in the nervous system is accomplished by rapid, synchronized firing of nerve cells in different parts of the brain. Just as rapid neuronal oscillations bind together different functional parts within the brain and nervous system, so rhythm binds together the individual nervous systems of a human community. [58]

As we saw in Chapter 8, Droit-Volet and Gil have pointed out that temporal distortions can be a sign of excellent response – they show our natural ability to adapt to environmental changes. Some of these distortions can be limiting, others have a balancing and healing effect. A revealing case of beneficial temporal contagion is the 'awakening' effect healthy rhythms can have on severely disabled people. An example of the non-beneficial type is the dancing plague which haunted Europe between 1200 an 1600. Individuals would fall into a trance-like state and danced for days and nights until they collapsed. There have been many attempts to explain the contagion. John Waller believes it was a kind of hysteria, a mass psychogenious illness, and that spontaneous trance is they key to understanding such a plague.

> The disruption of consciousness in the form of trance entails a number of cognitive changes. These include an impairment of critical analysis, reality checking, rational thought, a breakdown of the normal perception of space and time, an increase in suggestibility to internal or external stimuli, (...) and a heightened threshold for tolerating pain. The dancing mania of 1518 conforms closely to these criteria for trance. A distorted sense of time and a relative insensitivity to pain allowed the choreomaniacs to dance near-continuously for days or even weeks. [59]

Waller lists ways of inducing a trance: repetitive singing, chanting, drumming, dancing. He also mentions the possibility of narrowing one's range of LODs in order to reach this stage:

> Trance can also be achieved through becoming very highly focused on just one or two stimuli, disrupting the normal, broken flow of consciousness; this certainly accords with the yogic meditation practice of concentrating on a single thought or image and the use of a metronome or swinging to induce hypnosis. [60]

Contagious external stimuli can come in the shape of rhythms into which we inadvertently lock or nest. But also self-induced rhythmic activity can change our temporal perception, in particular, that of duration. I believe we tend to compensate the time dilations or contractions which occur in our everyday lives by increasing or reducing Δt_{depth}. Many of us have had the experience of falling into a little song – whistling, maybe, or humming a chant – to help us through a long or boring task. Are we, at such moments, trying unconsciously to spped up time by creating simultaneity? In other words, does our chanting (for example) provide a parallel level of description? Could it be so easy as to say that, if things get boring, just add Δt_{depth}? And, similarly, when we whistle in the dark during a potentially unpleasant experience, does this create the added value of an additional LOD, which shortens the experience?

Nesting and de-nesting – that is to say, increasing or reducing Δt_{depth} – require a range of possible responses to be at one's disposal. Entrainment is one such response – it is really a simulation of the outside world performed by our mirror neurons. If those neurons no longer perform, we cannot simulate our environment, which means that entrainment will no longer work as a means of re-activating learned behaviour or forming a new, healthy attractor. This is bad news for those American college students whose mirror neurons no longer fully respond. Should the trend continue and if any of them were to suffer the misfortune of becoming Parkinsonian patients, they may not be able to profit from the relief provided by simple spatio-temporal entrainment measures.

As is shown by the cases of temporal misfits presented in this chapter, successful embedding not only requires an endo-exo matching exercise but can only happen if the observer-participant has a sufficient choice of possible responses. This choice may be a range of rhythms which allow him to entrain into another person's dynamics and thus catapult himself onto other spatio-temporal LODs. An external stimulus which teaches new or re-kindles old behaviour can shift someone onto another attractor only if the dynamics of that external stimulus can be simulated by mirror neurons. Inactive mirror neurons manifest themselves in a loss of empathy – a state which would truly cut us off from the rest of the world.

Temporal misfits may be measured in terms of their degree of embeddedness – that is to say, their extension in Δt_{depth}. This includes their potential range of responses: their internal complexity as a measure of healthy dynamics, as suggested by West. This type of complexity can be assessed by measuring the individual's degree of fractality (the fractal dimension). A further measure to consider when determining the notion of a temporal misfit is the boundary complexity defined in Chapter 5 as the relation betweeen Δt_{depth} (endo) and Δt_{depth} (exo).

In the next chapter, we shall look at ideal rhythmic embedding and the connection between timelessness and duration. Although we all have a natural ability to detect such successful nesting – we can see poetry in motion – not all of the concepts introduced will be intuitive.

Fig. 9.1. "This taught me you can catch rhythm in photography." [61]

References

1. S.H. Konrath, *Empathy is Declining over Time in American College Students.* Presented at the Association for Psychological Science Annual Convention, Boston, May 27-30, 2010.
2. (no author), Empathy: College Students Don't Have as Much as They Used to, Study Finds. *Science Daily.* Retrieved June 3, 2010, from http://www.sciencedaily.com/releases/2010/05/100528081434.htm
3. T.J. Baudson, A. Seemüller and M. Dresler, *Chronobiology and Chronopsychology:* An Introduction, in: T.G. Baudson, A. Seemüller and M. Dresler (Eds.), *Chronobiology and Chronopsychology,* Pabst Science Publishers, Berlin 2009, p. 8.
4. T.G. Baudson, A. Seemüller and M. Dresler (Eds.), *Chronobiology and Chronopsychology,* Pabst Science Publishers, Berlin 2009.
5. J. Zulley, Human Chronobiology, in: T.G. Baudson, A. Seemüller and M. Dresler (Eds.), *Chronobiology and Chronopsychology,* Pabst Science Publishers, Berlin 2009, p. 14.
6. K. Brukamp, Mind: A-Head of Time, in: T.G. Baudson, A. Seemüller and M. Dresler (Eds.), *Chronobiology and Chronopsychology,* Pabst Science Publishers, Berlin, 2009, p. 37.
7. C.H. Cho and J.S. Takahashi, Molecular Components of the Mammalian Circadian Clock, in: *Human Molecular Genetics,* Vol. 15 (2006), pp. 271-277.
8. K. Brukamp, Mind: A-Head of Time, in: T.G. Baudson, A. Seemüller and M. Dresler (Eds.), *Chronobiology and Chronopsychology,* Pabst Science Publishers, Berlin 2009, pp. 36-47.
9. ibid, p. 38.
10. R.A. Wever, J. Polasek and N.C.M. Wildgruber, Bright Light Affects Human Circadian Rhythms, in: *Pflügers Archiv,* Vol. 391, pp. 85-87.
11. J. Zulley, Human Chronobiology, in: T.G. Baudson, A. Seemüller and M. Dresler (Eds.), *Chronobiology and Chronopsychology,* Pabst Science Publishers, Berlin 2009, p. 17.
12. Wikipedia: *Chronobiology,* http://en.wikipedia.org/wiki/Chronobiology, accessed 30.5.2010.
13. J. Zulley, Human Chronobiology, in: T.G. Baudson, A. Seemüller and M. Dresler (Eds.), *Chronobiology and Chronopsychology,* Pabst Science Publishers Berlin 2009, p. 18.
14. ibid.
15. C. Hahn, Human Circadian Rhythms, in: T.G. Baudson, A. Seemüller and M. Dresler (Eds.), *Chronobiology and Chronopsychology,* Pabst Science Publishers, Berlin 2009, p. 55.
16. G. Salvendy (Ed.), *Handbook of Human Factors and Ergonomics* (2nd edition), Wiley, New York 1997.

17. H. Lefebre, *Rhythmanalysis: Space, Time and Everyday Life*. Éditions Syllepse, Paris 1992 (English translation: Continuum, London 2004)
18. ibid, p. 75.
19. ibid, p. 80.
20. ibid, p. 81.
21. O. van Nieuwenhuijze, The Equation of Health, in: D.M. Dubois (Ed.) *International Journal of Computing Anticipatory Systems*, Vol. 22 (2008), CHAOS, Liège, pp. 235-249.
22. For a selection of O. v. Nieuwenhuijze's publications, see ottovannieuwenhuijze.googlepages.com and www.scienceoflife.nl
23. ibid, p. 244.
24. O. van Nieuwenhuijze, personal correspondence 2010.
25. M.C. Mackey and L. Glass, *Science* (1977) Vol. 197, p. 287.
26. G. Baier and T. Hermann, Temporal Perspective from Auditory Perception, in: S. Vrobel, O.E. Rössler, T. Marks-Tarlow (Eds.), *Simultaneity – Temporal Structures and Observer Perspectives*, World Scientific, Singapore 2008.
27. ibid, p. 349.
28. ibid, pp. 358-359.
29. ibid, pp. 360-361.
30. ibid, p. 353.
31. ibid, p. 362.
32. L. Nottale and P. Timar, Relativity of Scales: Application to an Endo-Perspective of Temporal Structures, in: S. Vrobel, O.E. Rössler, T. Marks-Tarlow (Eds.), *Simultaneity – Temporal Structures and Observer Perspectives*, World Scientific, Singapore 2008, pp. 229-242.
33. O. Sacks, *Awakenings*, New Revised Edition, Picador, London 1991.
34. ibid, p. 364.
35. ibid, p. 344.
36. L. Nottale and P. Timar, Relativity of Scales: Application to an Endo-Perspective of Temporal Structures, in: S. Vrobel, O.E. Rössler, T. Marks-Tarlow (Eds.), *Simultaneity – Temporal Structures and Observer Perspectives*, World Scientific, Singapore 2008, pp. 229-242.
37. ibid, p. 229.
38. ibid, p. 236.
39. ibid, p. 237.
40. O. Sacks, *Awakenings*, New Revised Edition, Picador, London 1991, p. 345.
41. B.J. West, Where Medicine Went Wrong: Rediscovering the Road to Complexity. World Scientific, Singapore 2006.
42. ibid, p. 303.
43. ibid, p. 305.
44. ibid, p. 307.
45. ibid, p. 285.

46. M.E. Morris, R. Iansek, J.J. Summers and T.A. Matyas, Motor Control Considerations for the Rehabilitation of Gait in Parkinson's Disease, in: *Advances in Psychology*, Vol. 111 (1995), pp. 61-93.
47. S. Hazard, Music Therapy in Parkinson's Disease, in: *Voices*, Vol. 8 (November 2008)
48. O. Sacks, Musicophilia. Tales of Music and the Brain. Picador, London 2008.
49. B.J. West, Where Medicine Went Wrong: Rediscovering the Road to Complexity. World Scientific, Singapore 2006, p. 289.
50. ibid, p. 219.
51. M.J. Kurz, K. Markopoulou and N. Stergiou, Attractor Divergence as a Metric for Assessing Walking Balance, in: *Nonlinear Dynamics, Psychology and the Life Sciences*, Vol. 14 (2010), No. 2, p. 161.
52. M. Koyama, T. Iokibe and T. Sugiura, Quantitative Symptom Discrimination of Parkinson's Disease by Chaotic Approach, in: IEICE Trans. Fundamentals, Vol. E83-A, No. 2, February 2000.
53. Scholarpedia: *Multi-Stability in Neuronal Models*, www.scholarpedia.org, accessed 27.6.2010.
54. Talk held by Bernd Ruf at the Free Waldorf School Wetterau, Bad Nauheim, Germany, June 2010.
55. (http://portal/unesco.org/education).
56. Ruf does not use the terms nesting and entrainment, he refers to the re-establishment of internal and external rhythms.
57. D. De Grave, From Insufficiency to Anticipation: an Introduction to 'Lichaamskaart', in: D.M. Dubois (Ed.) *International Journal of Computing Anticipatory Systems*, Vol. 22 (2008), CHAOS, Liège, pp. 145-254.
58. O. Sacks, *Musicophilia*, Picador, London 2008, p. 269.
59. J. Waller, A Time to Dance, a Time to Die. The Extraordinary Story of the Dancing Plague of 1518, Icon Books, London 2008.
60. ibid.
61. Photograph courtesy of Richard Thomson, 1985.

Chapter 10

Modifying Duration II:
Time Condensation

When the young protagonist in the rock opera *Tommy* develops into a Pinball Wizard, more than entrainment occurs: "He stands like a statue, becomes part of the machine, feeling all the bumpers, always playing clean, plays by intuition, the digit counters fall, that deaf and dumb and blind kid sure plays a mean pinball." [1] The young man in question has not simply locked into the rhythm of the machine but has experienced a boundary shift too. While locking into another person's stride or breathing rhythm can significantly change one's own wellbeing, reducing or extending one's boundaries is a yet more profound change. As we saw in Chapter 4, reduced or extended observer-participants have re-calibrated their body map and truly severed an internal or incorporated an external object or rhythm. Although many observer-participant extensions remain elusive (as is the case with Tommy becoming part of the machine), others can be pinned down by external observers. Extensions can be made transparent if our internal range of possible responses to environmental changes allows us, for instance, to make up for delays by means of anticipatory regulation. In the spatial domain, we may incorporate a tool to such an extent that our brains remap, for instance, far space as near space (see Chapter 4).

In Chapter 1, we saw how time slows down for people who are "in the zone" – that is to say, people who are totally immersed and focussed. They have significantly reduced Δt_{depth} through shifting their attention to one LOD and thus extended Δt_{length}, accommodating far more detail on one LOD than during a normal distribution of Δt_{depth} and Δt_{length}. So was Tommy the Pinball Wizard "in the zone"?

Natasha Schull has studied the behaviour of individuals who are addicted to modern sophisticated gambling machines and came to the conclusion that it is not the desire to win which motivates addicts, but the experience of being "in the zone". Although winning might be part of the initial hook, according to Schull, it fades soon, whereas the allure of the zone does not. The zone is

> a dissociative state or trance in which players lose a sense of time, space and physical embodiment, consumed totally by the spinning numbers, symbols or electronic card hands before their eyes. (...) players describe the zone as a compelling, mesmerizing condition of intense concentration – an almost out-of-body experience. Heavy machine gamblers come to crave this state. [2]

Admittedly, there is a difference between a pinball machine and a piece of modern gambling equipment. But, for most players, focussing and concentration – that is to say, a reduction of Δt_{depth} – are required in the handling of both of them. We can assume, though, that Tommy was no longer "in the zone" or, as we have also termed it, in a state of "flow". He had surpassed that stage and actually formed a new systemic whole by re-calibrating his body map so as to assimilate the dynamics of the pinball machine. The conditioning effect of playing the machine is that anticipation and the incorporation of delay times re-calibrate the system boundaries: the player becomes a temporally extended observer-participant.

An even more intriguing kind of temporal distortion occurs when the LODs of a self-similar nesting cascade can be translated into each other effortlessly. In an idealized mathematical model, Δt_{depth} would, at that point, approach infinity and Δt_{length} would approach zero. Such translation is only possible when the temporal structures on the nested LODs are self-similar – that is to say, if they only vary in scale. We have briefly looked at an example of such a temporal self-similar fractal in Chapters 2 and 3, when we described the phenomenon of overtones and the missing fundamental. As the overtones are integer multiples of the fundamental frequency, they are easily translatable into each other – in fact, they are scale-invariant. This scaling generates a superposition of harmonics, embedding frequencies of different lengths.

In this chapter, we shall look at the phenomenon of insight – the instantaneous realization of a highly complex idea – as a manifestation of time condensation, which occurs when Δt_{depth} approaches ∞ and Δt_{length} approaches 0 [3]. This phenomenon is a state of zero temporal resistance in the dimension of Δt_{length} and, therefore, the experience of instantaneousness – that is to say, sheer simultaneity. While being "in the zone" or flow occurs as a result of reducing Δt_{depth} and expanding Δt_{length}, insight and condensation are triggered by the opposite mechanism (see Table 10.1).

To start with, however, let us look at some of the phenomena which make time slow down and which we have already encountered. I have suggested that the common denominator in these temporal distortions is de-nesting – that is to say, the reduction in Δt_{depth} [4]. First of all, there are stressful situations, when we de-contextualize in order to be able to focus on detail. As we have seen in Chapter 1, accounts of narrowed visual and auditory perception are often accompanied by subjective time dilation. When we shut out visual or auditory stimuli, we diminish the number of simultaneous perceptions and thus reduce Δt_{depth}.

Table 10.1. The distribution of Δt_{depth} and Δt_{length} invokes either flow or insight.

	distribution of Δt_{depth} and Δt_{length}	
- Focussing in stressful situations - flow / being "in the zone"	reducing Δt_{depth}	increasing Δt_{length}
- insight (condensation)	increasing Δt_{depth}	reducing Δt_{length}

Another example to be listed in this context is de-nesting by means of removing delays. In Chapter 4, we saw how the order of events is distorted for us if a delay we have grown accustomed to is suddenly removed. Delay times are temporal observer-participant extensions which widen the protension part of our Now – that is to say, the part which anticipates the future. This means that if a delay is shortened or removed, less nesting and thus less simultaneity (Δt_{depth}) exists within the Now. Another important way of de-nesting is by means of synchronization. Entrainment is a reduction in Δt_{depth}, as all oscillations are simultaneous. Nested events, by contrast, overlap and cannot be reduced to one LOD (unless they are self-similar, as we shall see later in this chapter). As we saw in Chapter 6, the synchronization of gamma waves in the brain has been correlated with transcendental meditative states – the "frequency of silence" sought by meditating monks when they focus, for instance, on a mantra in order to be rid of all other thought. During such synchronization, stimuli are de-nested, which means that there is a reduction in Δt_{depth}. The same mechanism is at work when our body locks into external rhythms – that is to say, during entrainment between inside and outside. Next, we saw in Chapter 7 how diminishing contrast correlates with the act of de-nesting and the perception of time slowing down. The amount of contrast we perceive determines the speed which embedded objects seem to have. A higher contrast generates more context, so embedded objects appear to speed up. Conversely, a lower contrast creates less context and slows down embedded objects. And, finally, also in Chapter 7, de-nesting was shown to be triggered by a reduction in complexity. This can be achieved by grouping percepts into gestalts, thus reducing Δt_{depth}. To sum up:

De-nesting is attained by means of

- ... inducing stress;
- ... shortening or removing a delay;
- ... synchronization (entrainment);
- ... diminishing contrast;
- ... reducing complexity. (However, this is also a trademark of insight and condensation, as we shall see.)

By contrast, nesting is attained through

- ... learning and training (spatio-temporal contextualization);
- ... increasing internal complexity (i.e. the range of possible responses).

Having called back to mind a number of conditions which accompany a reduction in Δt_{depth}, let us now look at situations where a sudden collapse of nested LODs in the dimension of Δt_{depth} leads to insight via condensation.

The term "insight" has as many meanings as there are disciplines. Let us refer to it here as the instantaneous realization of a complex issue by means of complexity reduction. The notion of insight is related to the Greek concept of *noesis*, suggesting an immediate and embodied understanding. "Insight" has often been loosely paraphrased as 'epiphany' or 'the Eureka moment' – that is, a sudden recognition of cause and effect, or an intuitive solution which seems to have come unexpectedly, as if from nowhere. However, insight does not fall from the sky but is always the result of training and of incorporated knowledge – knowledge which may pop up in a flash, in condensed form, and uninvited [5].

I have found Roger Penrose's notion of insight very attractive because it describes a non-algorithmic, instantaneous event. Being non-algorithmic, it cannot have any extension in Δt_{length} but only in Δt_{depth}. Penrose illustrates the idea of a non-temporal experience of insight with Mozart's ability to perceive highly condensed structures:

An extreme example (...) is Mozart's ability to 'seize at a glance' an entire musical composition 'though it may be long'. One must assume, from Mozart's description, that this 'glance' contains the essentials of the entire composition, yet that the actual external time span, in ordinary physical terms, of this conscious act of perception, could be in no way comparable with the time that the composition would take to perform. [6]

Another composer cited by Penrose is J. S. Bach, who must have organized his compositional work in such a way that it is to a certain extent self-similar: the character of an entire composition can be anticipated in the shortest sections. However, in order to appreciate and anticipate the wholeness and complexity of the composition, the listener's fractal temporal interface must be sufficiently differentiated – that is to say, he must have acquired an adequate number of LODs, which manifest themselves as possible responses.

> Listen to the quadruple fugue in the final part of J.S. Bach's Art of Fugue. No-one with a feeling for Bach's music can help being moved as the composition stops after ten minutes of performance, just after the third theme enters. The composition as a whole still seems somehow to be 'there', but now it has faded from us in an instant. Bach died before he was able to complete his work and his musical score simply stops at that point, with no written indication as to how he intended to continue. Yet it starts with such an assurance and total mastery that one cannot imagine that Bach did not hold the essentials of the entire composition in his head at the time. [7]

Penrose knew what he was talking about: he describes a flash-like insight he had while crossing a street and being preoccupied with a different matter. Suddenly, the solution to a problem – concerning the point of no return during the collapse of black holes – occurred to him in a flash. Penrose stresses the non-temporal, non-algorithmic character of this insight: no calculation, no thinking (which would have taken time) happened when it struck – in other words: there was no Δt_{length}.

For Penrose, the underlying motivation for these considerations was his aim to form a bridge between the quantum world and the classical one, including relativity. In his theory-to-be of correct quantum gravity, non-algorithmic elements on the quantum level are catapulted onto the macroscopic level of consciousness. It is somewhere along the lines of this event that, according to Penrose, the arrow of time slips in. The

R-Part of quantum theory, which corresponds to the wave function collapse, contains those non-algorithmic, catapulted elements (in the wake of which the arrow of time appears in the macroscopic world). He is aiming for a physics of the mind, in which human consciousness acts as the pivotal point between the physical world of algorithms and Plato's realm of mathematical, timeless ideas. Contact with that world can only occur as insight – a non-temporal access to a timeless world. By contrast, our conscious connection to the physical world is algorithmic – it takes time to compute.

In the language of my Theory of Fractal Time, insight corresponds to condensation: Δt_{length} approaches zero, whereas Δt_{depth} approaches ∞ and then collapses. Having discussed Penrose's notion of human insight, let us now consider a different, mathematical model of this concept, exemplified by the condensation velocity of two ideal fractal structures.

Elsewhere, I have defined condensation as a property generated by temporally congruent nestings within one observer perspective [8]. It is measured in the quantities of condensation velocity $v(c)$ and condensation acceleration $a(c)$, where $v(c)$ and $a(c)$ are derived from Δt_{depth} and Δt_{length}. The quotient of Δt_{length} of LOD1 and Δt_{length} of LOD2 equals the condensation velocity $v(c)$ for LOD2 $\sim$ LOD1 ($\sim$ denotes *nested in*). For scale-invariant structures – that is to say, self-similar fractals – $v(c)$ is identical with the scaling factor (s). Δt_{depth} comes into the equation as the number of nested LODs taken into account.

These notions are exemplified below by a mathematical fractal, the Koch curve, which we have already encountered in Chapter 2. For all scale-invariant mathematical fractals, the condensation velocity is identical to the scaling factor.

The condensation acceleration $a(c)$ equals 1 in mathematical structures, as the scaling factor is constant. So, for the Koch curve, the condensation velocity $v(c) = 3$ and the condensation acceleration $v(a) = 1$.

The Cantor dust is another fractal which displays a condensation velocity of 3 and, as it also scales at regular intervals, the condensation acceleration is also 1.

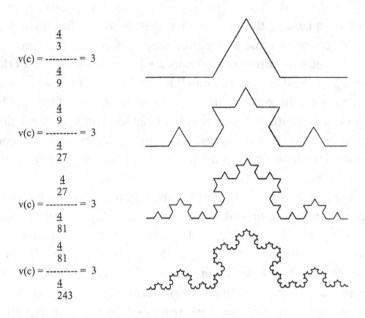

$$v(c) = \frac{\frac{4}{3}}{\frac{4}{9}} = 3$$

$$v(c) = \frac{\frac{4}{9}}{\frac{4}{27}} = 3$$

$$v(c) = \frac{\frac{4}{27}}{\frac{4}{81}} = 3$$

$$v(c) = \frac{\frac{4}{81}}{\frac{4}{243}} = 3$$

Fig. 10.1. The condensation velocity of the Koch curve. [9]

It is of no importance whether the fractal dimension lies between 0 and 1 (as in the generation of the Cantor dust), between 1 and 2 (as for the long line of the Koch curve), between 2 or 3 (as for the Menger sponge) or is of any other dimensionality. The scaling factor alone determines the condensation scenario.

For natural fractals, which display an upper and a lower limit to their scale invariance, the condensation velocity and acceleration need to be determined individually for each nesting and embedding LOD.

If long and short time intervals on nested LODs exhibit the same pattern of change, for instance, the same variability or fluctuations on all levels, we may define this shared pattern – be it statistical or not – as a constant which serves as a reference scale in order to relate the intervals on the nested LODs. If we are dealing with non-statistical data – that is to say, if we can define nested LODs by unit intervals, a smallest possible interval can be determined. In an earlier work, I defined an interval which is indivisible (in the sense that it cannot host further nestings) as the Prime.

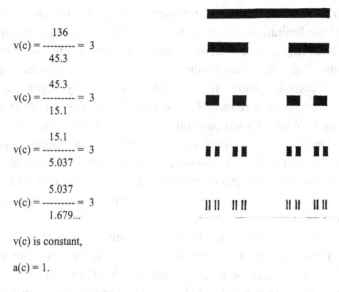

$$v(c) = \frac{136}{45.3} = 3$$

$$v(c) = \frac{45.3}{15.1} = 3$$

$$v(c) = \frac{15.1}{5.037} = 3$$

$$v(c) = \frac{5.037}{1.679...} = 3$$

v(c) is constant,

a(c) = 1.

Fig. 10.2. The condensation velocity of the Cantor dust. [10]

The Prime is a temporal natural constraint (TNC), since it acts as a limitation on temporal structurability. It is the smallest interval within a nesting cascade and indivisible in the Bergsonian sense. No Prime can be defined for statistical scale-invariance. No limit is defined for the uppermost LOD – that is to say, the most extended interval which embeds all others. This is so because our fractal temporal perspective is a phenomenological model which describes the observer-participant's nested Nows. As long as he is capable of nesting existing Nows into more current ones, there is no upper limit for the nesting cascade. In his Theory of Scale Relativity, Nottale has defined not only a smallest length scale, but also a largest one: he equates the smallest, indivisible interval with the Planck scale and the uppermost LOD with the cosmic length scale which is related to the cosmological constant [11]. In my theory, the uppermost 'length scale' is our Now – our only interface with the world.

TNCs on possible embeddings can be generated both by external, physical structures and the observer-participant's internal temporal differentiation – that is to say, his degree of complexity measured in the number of possible alternative responses: Δt_{depth}. As we saw in Chapter 3, some of the limitations to our experience of the two temporal dimensions Δt_{depth} and Δt_{length} result from constraints imposed by our perceptual apparatus. But rather than seeing TNCs as a limiting nuisance, it makes sense to consider them as a selection effect. By setting a limited reference scale, they reduce the level of complexity, as arbitrariness is eliminated. Without such constraints, no boundaries would be definable within which LODs and units may be arranged. Our primary experiences of time (simultaneity, succession, duration and the Now) also act as TNCs on our temporal fractal perspectives. And thus, TNCs can be seen as facilitators: they render possible communication, travel, trade and science [12].

Reduction in complexity is usually a telltale sign of a decrease in Δt_{depth}. However, for scale-invariant structures, the opposite happens: first, there is an increase in Δt_{depth} and then a collapse of all LODs onto one level. Synchronization – that is to say, entrainment – can occur after a scale transformation of space-time on all LODs. The common denominator which translates between the scales is the structure of the Prime which recurs on all LODs. In addition, gestalts and metaphors may also act as TNCs, as we generate and perceive them as indivisible wholes. We form such entities pre-attentively into indivisible wholes which are not accessible to introspective analysis [13]. So, we are not aware of many a cause of complexity reduction we take advantage of. This is also true of condensation.

In order to properly describe a condensation scenario, it is necessary to introduce yet another quantity: the Prime structure constant (PSC). If the structure of a Prime – that is to say, the scaling structure which recurs on all nested LODs – is set as a constant, a kind of scale relativity emerges: space-time is "bent" with respect to the PSC. This can only happen if there is structural congruence (after a scale transformation) on all or a number of LODs.

If this congruence is achieved by a sentient being – for instance, a human observer-participant – who registers only the PSC, thus disregarding the length of the individual nested scaling structures, condensation sets in. Condensation is sheer simultaneity within a fractal temporal perspective: it is instantaneous, as it creates no friction in

Δt_{length}. A helpful image is a kind of "vertical superposition", in which the entire nesting cascade behaves like one LOD. (vaguely reminiscent of a Bose-Einstein condensate). On the phenomenological level, condensation manifests itself in the experience of insight.

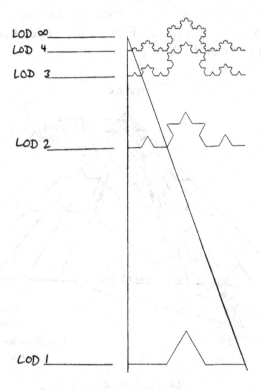

Fig. 10.3. The PSC of the Koch curve.

Long ago, when I read about Penrose's idea of a non-algorithmic access to Plato's world of ideas, I was inspired to develop a theory of fractal time which endeavours to connect our temporal world with the non-temporal realm of the Primes. A complete illustration of the mechanisms at work during a condensation scenario is shown in Fig. 10.4.

Condensation requires preparation. The individual LODs which are to collapse into one must first come into existence. This happens via perceptual and cognitive experience and is a learning process which takes time: intuition does not appear from out of the blue but is the result of thorough training and embodied memory [14]. At this stage, we create both Δt_{length} and Δt_{depth}. With every new learning step – that is to say, with every contextualization – we embed more LODs in the nesting cascade.

At the same time, Δt_{length} contracts further and further while Δt_{depth} increases. At the limit, Δt_{length} approaches 0 and Δt_{depth} approaches ∞. This is the point of direct contact between the world of duration and the non-temporal realm of the Primes.

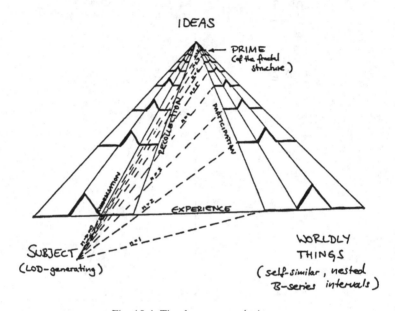

Fig. 10.4. Timeless access via Δt_{depth}.

Figure 10.4 shows this relation in the shape of Plato's Triangle, in which I have replaced the original term 'anamnesis' with 'recollection'. It is simply the nested retentions and protensions within the Now – that is to say, our memory of the past and the future – which are represented on the right side of the triangle. The fractal notion does not imply aspects such as the cleansing of the soul from the encrustation of bodily cares and interest, as does the term anamnesis. In the fractal model, Nature's participation in ideas manifests itself exclusively in structural congruence via the PSC, which displays characteristics of both the realm of worldly things and that of ideas. Whereas, in the realm of worldly things, the structure of the Prime extends in both Δt_{length} and Δt_{depth}, in the world of ideas, it corresponds to the last nesting, which is no longer divisible and thus timeless. At this point, the structure of the Prime has lost its nesting potential – that is to say, it can generate no further Δt_{depth} – and has become timeless in both temporal dimensions. The smallest interval of the nesting cascade thus operates as a TNC which forms the boundary between time and timelessness by condensing into a state of indivisibility.

The observer-participant needs to have built up a fractal temporal perspective with sufficient Δt_{depth} to initiate or partake in a condensation scenario. Condensation itself – the Eureka moment or insight – is sheer simultaneity. The cognitive work which follows in the reflection of this experience again generates Δt_{length}. I have determined four conditions required for condensation to take place:

(1) A nesting of temporal intervals must be recognized or generated;
(2) This nesting must exhibit scale-invariance, preferably in the form of the PSC recurring on all LODs available;
(3) The scale-invariance must be other than statistical. Statistical scale-invariance does not suffice, since a statistical determination always rests on a large amount of data and therefore implies divisible extension. It is not possible to define a Prime for statistical scale-invariance;
(4) The observer-participant's temporal perspective – that is to say, his Now – must be an extended present in Husserl's sense, which comprises the basic structure of the Prime of the fractal in question. [15]

Sheer simultaneity is not only experienced in insight which rests on training and knowledge. Embodied insight is often sought in states of trance by building up a nesting cascade of synchronous movement of all parts of the body and then collapsing it into one LOD. Such physical synchronization is often accompanied by a cognitive effort in the form of focussing, for instance, on a spiritual theme. The famous Whirling Dervishes of Konya perform their religious dance and prayer ceremony, the sema, in which they reach a state of ecstasy by spinning on their right foot. At the same time, they stretch out their right palm towards heaven and their left towards the ground. Rotating around their own centres, they create a synchronous movement in which their outer parts spin faster than their inner ones – their arms faster than their hips, their hips faster than their coccyx. The closer to the vertical spinning axis the body parts are located, the slower the rotation until, at the limit in the centre, the spinning occurs infinitely slowly. (One may speculate, in the same vein, whether Tibetan prayer wheels are outsourced focussing of diminished observer-participants or incorporated tools of extended ones).

In his *Four Quartets*, T. S. Eliot beautifully captured this idea of an infinite amount of time in Δt_{depth} but none in Δt_{length}:

> At the still point of the turning world. Neither flesh nor fleshless;
> Neither from nor towards; at the still point, there the dance is,
> But neither arrest nor movement. And do not call it fixity,
> Where past and future are gathered. Neither movement from nor
>   towards,
> Neither ascent nor decline. Except for the point, the still point,
> There would be no dance, and there is only the dance.
> I can only say, *there* we have been: but I cannot say where.
> And I cannot say, how long, for that is to place it in time. [16]

Approaching a state of infinite Δt_{depth} will, at first, increase complexity, as we are learning and thereby creating new LODs. The more LODs are nested, the more likely we are to form gestalts and metaphors, which condense a number of LODs into one (as in Kohler's experiment outlined in Chapter 7). For a condensation scenario to take place, an extensive nesting cascade must have been built up which may then collapse onto one LOD. The collapse will then inevitably lead to a reduction of the level of complexity of a system, as we grasp everything in an instant.

German sociologist Niklas Luhmann, founder of sociological systems theory, sees trust as being our most powerful means for reducing the complexity of the world in which we are embedded [17]. Without trust, we would not dare to leave our bed in the morning. If anything is possible, the individual is faced with an unmanageable level of complexity. Initially, the endo-perspective is incredibly complex, according to Luhmann. Trust is formed in an interactive field which is influenced both by psychological and social formations of systems. This interactive field is reminiscent of Winnicott's potential spaces (see Chapter 5), which are likewise characterized by the fact that they cannot be assigned either to the inside or the outside. Luhmann defines complexity only in terms of the difference between system and environment and the system's potential to evolve. Complexity in functionalist systems theory refers to the number of options rendered possible by the formation of systems. In cybernetic systems theory, on the other hand, the same relationship between system and world appears as the establishment of a higher order. This order is characterized by a lower level of complexity as a result of the formation of systems and can be seen as a selection effect. For all sentient beings, the world is overpoweringly complex because it provides more possibilities than the system can react to in order to attain or maintain homeostasis. A human being can grasp mere possibilities because he experiences others who, at that moment, experience what to him is only a possibility. At the same time, he experiences being perceived by others as an object. This way, he can adopt their perspectives and can identify himself. Thus, other human beings act as mediators of identity and the world's complexity. Luhmann introduces a novel dimension of complexity, namely,

> that of the experienced (perceived) and interpreted subjective I-ness of the other human being. The other human being has an original access to the world, could experience everything differently from the way I do and can thus radically throw me into doubt. (...) The complexity of the world is (thus) extended again through the social dimension, which bring the other human being into my consciousness not only as an object but as another I.

Therefore we need, with this additional complication, at the same time novel mechanisms for reducing complexity (...) [18]

Trust is a very powerful mechanism for reducing complexity. It is a temporal observer extension, as it anticipates the future. By trusting, we exclude certain developments from the range of possibilities. We base our decisions – which possibilities to take into account and which ones to discard – on a strong belief in continuity. We expect that our fellow human beings will act tomorrow roughly the way they did today and yesterday, so we can leave the house unarmed and drive at a convenient speed. Co-ordinatory action and co-operative behaviour – which grow in the wake of trust – reduce the complexity of the world and make it easier for us to navigate successfully.

But trust is anticipation – we assign the future to our Now by allowing protension to influence our decisions. However, we will only know in retrospect whether our trust was justified or not. But our decision about whether or not to trust has to be made in the Now. Luhmann stresses that trust always means taking a risk, is always unaccountable. In Luhmann's view, even if most of us would be able to come up with countless reasons to justify our decision to trust in this or that case, those reasons serve mainly as a social justification and self-assurance. This is why decision theory acknowledges that the time factor introduced by anticipation cannot simply be equated with an uncertainty factor. At the point a decision has to be made, knowledge which is not accessible to the decision maker is also not available to him in terms of certainty and uncertainty. So whether we employ someone or make a financial investment, we build a bridge of trust to serve us until we can have certainty.

The problem complexity poses is thus distributed and reduced. We trust each other for the time being to be able to manage unclear situations successfully and thereby reduce complexity. On the basis of this trust, the other person then has a better chance of actually being successful [19].

Luhmann's notion of reduction cannot be equated with deduction. It is, rather, induction, as trust is, in the end, always unaccountable. In terms of a fractal perspective, a reduction of complexity would result first from an increase in Δt_{depth}, as more simultaneous LODs are present which allow the formation of gestalts. This would go hand in hand with the formation of trust. When a certain degree of Δt_{depth} has been reached at some saturation point, condensation may set in in the shape of a collapse of all LODs into one. This scenario would manifest itself as oceanic boundary loss – that is to say, a state of total immersion (see Metzinger's accounts in Chapter 5) and infinite trust. The advantages of such a state would be smooth and efficient behaviour, allowing almost frictionless navigation, like superfluid agents would experience in a superfluid environment. The disadvantage would be a loss of perspective and self – the transition from an observer-participant to a pure participant.

I have pointed out the correlations and shared characteristics between condensation and a reduction in complexity. Trust can be seen as the ultimate observer-participant extension, as it embraces and incorporates the future into the present, thus spanning a large interval without causing friction within the Now. A maximum outward boundary shift occurs which, in extreme cases, would result in oceanic boundary loss. So, in a way, the anticipation of a large interval is condensed into the present moment. This, however, is a condensation scenario which differs from that manifesting itself as insight. This insight is caused by the collapse of a large number of nested LODs onto a single one and has no room for anticipation.

As we have implied, the notion of intuition shares many characteristics with that of insight. The idea that intuition can happen in (almost) a flash has been greatly reinforced by Malcolm Gladwell's book *Blink!* [20]. Gladwell describes how our snap judgements are often far more effective than thorough analysis. He is, of course, aware that intuitive decisions can also be biased or primed. But that is not Gladwell's point. It is, rather, that we can learn to grasp a complex event within a glimpse and thus simplify our lives by a technique called thin-slicing. This is "the ability of our unconsciousness to find patterns in situations and behaviour based on very narrow slices of experience" [21]. The idea is that behaviour which we observe, say, during a decade, is also present in a 15-minute (or sometimes a 1-minute) time slice.

Although Gladwell does not use the word, he actually describes a self-similar temporal fractal. The essential information about the whole is present in all of its parts.

He also describes how a marriage consultant (and the individuals he had trained) were able to judge within three minutes whether a couple would stay together for the next 15 years. The consultant's method was to look for such vital cues as gestures of contempt and avoid being distracted by a polyphony of complex emotions. While such skills can be trained, others are harder to attain and need, indeed, a lifetime of experience. A former tennis player reports that, after watching someone's first serve, he could almost always tell when a double fault would occur. But he never found out exactly what triggered this certainty in him. It was not one single cue but rather a diffuse feeling of certainty. The point is that he acquired it in an instant – in the wink of an eye. A similar anecdote describes art experts who could tell at a glance that a statue was fake, even after a museum had spent a fortune and 14 months carrying out high-tech analyses, which had led them conclude that the artefact was genuine. The experts in question simply 'knew' instantly, although they could not justify this in any academic manner. One of them explained simply that his first impression was that it was 'fresh' – an adjective one would hardly associate with an ancient statue.

These intuitions came flash-like, immediately at the first encounter. We all know how important first impressions are, whether we are preparing for a job interview or a date. In such situations, it seems that we also reveal the same vital cues on all time scales, so that a glimpse or a very short time span suffices for a good evaluation. The success of speed-dating is based on our ability to grasp the essential cues within five minutes. That is to say, we would not be much wiser if we spent five days with the individual concerned.

How much complexity reduction we can take and still grasp the essential message is nicely illustrated by Gladwell's case studies concerning surgeons and their patients. It was found that a psychologist was able to assess the risk of malpractice lawsuits by using distorted recordings, as follows:

The psychologist Nalini Ambady listened to Levinson's tapes, zeroing in on the conversations that had been recorded between just surgeons and their patients. For each surgeon, she picked two patient conversations. Then, from each conversation, she selected two ten-second clips of the talking doctor, so her slice was a total of forty seconds. Finally, she 'content-filtered' the slices, which means she removed the high-frequency sounds from speech that enable us to recognize individual words. What's left after content-filtering is a kind of garble that preserves intonation, pitch, and rhythm but erases content. Using that slice – and that slice alone – Ambady (...) had judges rate the slices of garble for such qualities as warmth, hostility, dominance, and anxiousness, and she found that by those ratings, she could predict which surgeons got sued and which one's didn't. [22]

The results showed no correlation with the surgeon's experience or degree of skill. The conclusion to be drawn was that patients do not sue doctors they like – that is to say, doctors who listen (on average little more than 3 minutes longer than their less-liked colleagues), show concern and do not display dominant behaviour.

Complexity reduction can be trained. When we learn metaphors or form gestalts, we reduce some of the simultaneous LODs in our Now to a smaller number by connecting them with an internal logic that preserves the detail of the parts. We do not leap from the careful assembly of LODs we have nested through learning directly to condensation. The intermediate states allow partial collapse in order to reduce the complexity of the world. As a result, we facilitate our perceptions, by forming gestalts, for instance. At that stage, we continue to build up trust by reducing social complexity. These simplifications go hand-in-hand with a reduction of Δt_{depth}. Table 10.2. shows how three changes in our fractal temporal perspectives – that is to say, from an increase in complexity via complexity reduction to complexity compression – manifest themselves in our perception of time, our level of skill and involvement.

Note that, while focussing as a result of stressful behaviour, as well as flow or being "in the zone" reduces Δt_{depth} and, at the same time, increases Δt_{length}, there is no increase in Δt_{length}, in intuition or insight.

The earlier chapters of this book have dealt with the experience of time slowing down as a result of a reduction in Δt_{depth}. In this chapter, we have encountered a reduction in Δt_{depth} which is accompanied, not by time dilation, but by its opposite: time condensation. Whereas in stressful situations, we focus on one or few LODs, and in flow, we spread ourselves out on all LODs, insight has no extension in Δt_{length}. It manifests itself in an infinite increase in Δt_{depth} and subsequent collapse of LODs, without creating Δt_{length}. It is only in the phenomenon of insight as described in my condensation model, that a reduction of Δt_{depth} is not compensated by an increase in Δt_{length}. Insight is genuine time compression, non-temporal and non-local, which allows for instantaneous accessing of the Primes. In a condensation scenario, the temporal and non-temporal realms are reconciled.

Table 10.2. The correlates of complexity increase, decrease and compression.

$+ \Delta t_{depth}$	$- \Delta t_{depth}$	condensation
Increase in complexity	Complexity reduction	Complexity compression
Initial stages of learning, training, boundary formation	Advanced stages of learning, training, learning metaphors, forming gestalts, focussing, boundary shifts	Instantaneous access to knowledge, spontaneous event
Perceived time speeds up	Perceived time slows down	No duration
Learning, defining objects	Entrainment, flow, being "in the zone"	Insight, (intuition)
Initial stages of building up trust	Advanced stages of building up trust, anticipation, embodied trust, total immersion	No anticipation
More LODs Nesting, acquiring LODs	Fewer LODs De-nesting, condensing LODs by forming gestalts	Collapsed LODs Access to non-temporal Primes via the PSC

In the next chapter, we will take a closer look at the complex notion of the Now, including its role in the changing physical paradigms and philosophical interpretations. We will focus, in particular, on its function as an interface between the self and the world.

References

1. P. Townshend, *Pinball Wizard, Tommy, A Rock Opera*. Track, Polydor, UK 1969.
2. N. Schull, *Machine Zone: Technology Design and Gambling Addiction in Las Vegas*, Princeton University Press 2010. (MIT book announcement on www.physorg.com/print162052637.html).
3. S. Vrobel, *Fractal Time*, IAIR, Houston 1998.
4. S. Vrobel, *When Time Slows Down: The Joys and Woes of De-Nesting*, Talk and paper at InterSymp 2010, Baden-Baden, Germany.
5. M. Gladwell, *Blink – The Power of Thinking Without Thinking*. Penguin, London 2006.
6. R. Penrose, *The Emperor's New Mind*. Vintage, London 1989, p. 575.
7. ibid, p. 576.
8. S. Vrobel, *Fractal Time*, IAIR, Houston 1998.
9. ibid.
10. ibid.
11. L. Nottale, Scale Relativity, Fractal Space-Time and Morphogenesis of Structures, in: H.H. Diebner, T. Druckrey and P. Weibel (Eds.), *Sciences of the Interface*. Genista, Tübingen 2001, pp. 38-51.
12. S. Vrobel, Fractal Time and the Gift of Natural Constaints, in: *Tempos in Science and Nature: Structures, Relations, Complexity*. Annals of the New York Academy of Sciences Vol. 879 (1999), pp. 172-179.
13. S. Vrobel, Simultaneity and Contextualization: The Now's Fractal Event Horizon. Talk held at *13th Herbstakademie: Cognition and Embodiment*, Monte Verita, Ascona, Switzerland 2006.
14. M. Gladwell, *Blink – The Power of Thinking Without Thinking*. Penguin, London 2006.
15. S. Vrobel, *Fractal Time*, IAIR, Houston 1998.
16. T.S. Eliot, *Four Quartets*, Faber and Faber, London 1986, p. 15 (first published 1944).
17. N. Luhmann, *Vertrauen*, 4th edition, Lucius & Lucius, Stuttgart 2005 (first published 1968).
18. ibid, p. 6 (my translation).

19. ibid, p. 31.
20. M. Gladwell, *Blink – The Power of Thinking Without Thinking*. Penguin, London 2006.
21. ibid, p. 23.
22. ibid, p. 42.

Chapter 11

Defining Boundaries:
Why is it Always Now?

How could we explain our Now to an extraterrestrian? Billy Pilgrim, the involuntary time traveller in Kurt Vonnegut's *Slaughterhouse-Five* had the privilege of hearing an attempted explanation by the Tralfamadorians, who had kidnapped and taken him to their planet. At the zoo where Billy had been put on display, the guide tried to explain to the Tralfamadorian crowd what time looked like to an Earthling:

> The guide invited the crowd to imagine that they were looking across a desert at a mountain range on a day that was twinkling bright and clear. They could look at a peak or a bird or a cloud, at a stone right in front of them, or even down into a canyon behind them. But among them was this poor Earthling, and his head was encased in a steel sphere which he could never take off. There was only one eyehole through which he could look, and welded to that eyehole were six feet of pipe. This was only the beginning of Billy's miseries in the metaphor. He was also strapped to a steel lattice which was bolted to a flatcar on rails, and there was no way he could turn his head or touch the pipe. The far end of the pipe rested on a bi-pod which was also bolted to the flatcar. All Billy could see was the little dot at the end of the pipe. [1]

This was not the end of limitations to the human condition. Billy wasn't even aware of the fact that he was strapped to a flatcar. Neither was he able to realize that this vehicle sometimes moved fast, sometimes slow, accelerated and sometimes drew to a halt. He was blissfully unaware of everything the guide had described except for the little dot he could make out though the pipe. The Tralfamadorians could see in four dimensions. They were able to look at any moment and know that it is permanent: All Nows had always existed and would always do so. It was only poor Earthlings who thought that, once a moment has passed, it was irretrievably gone. Also the succession of Nows was an illusion only Earthlings fell for. When Tralfamadorians read telegrams or books, they read them all at once. Their books had no beginning, middle or end and it was the simultaneity of many marvellous Nows which created the beauty Earthlings could not appreciate. When Tralfamadorians looked at the sky, they didn't see little dots of light but a firmament filled with luminous spaghetti. They were aware that they lived in a totally deterministic universe and saw themselves as bugs in amber.

The Tralfamadorian model of the universe was actually a Minkowski space. In 1908, German mathematician Hermann Minkowski conceived a framework which had three spacelike dimensions and one timelike dimension. This was new – Euclidean space had only spacelike dimensions. After Einstein had published his Special Theory of Relativity, Minkowski searched for a way of representing in diagrammatic form the idea that space and time are inextricably intertwined, so that a union of the two was required to represent reality independently. A Minkowski diagram shows two observer's coordinate systems, superimposed, graphically representing different simultaneity horizons for both observers, as well as relativistic effects such as time dilation and length contraction. Minkowski spacetime is a four-dimensional manifold, in which each point is fixed in a position in space and time and represents an event. The succession of events form a world line (similar to the spaghetti-like shape the Tralfamadorians saw when they looked at the stars). Minkowski space (resembling a slab of jello) is penetrated by the trajectories of events that show as world lines. The jello is rather stiff, though. For, as Wheeler has pointed out, "spacetime does not wiggle. It is 3-D space geometry that undergoes agitation. The history of its wriggling registers itself in frozen form as spacetime." [2]

But let us go back to an earlier point in history, and to a similar idea of a container-like spacetime. Newton thought of time as being absolute (rather than relative). His container time did not have room for an observer perspective – there is no Now in Newton's model:

Absolute, true and mathematical time, of itself, flows equably without relation to anything external, and by another name is called duration. (...) For times and spaces are, as it were, the places as well of themselves as of all other things. All things are placed in time as to order of succession; and in space as to order of situation. [3]

By contrast, Leibniz' spatial and temporal extensions referred to the objects themselves, rather than to some absolute medium within which motion takes place, as Newton claimed. For Leibniz, time and space were just convenient ways of measuring, so it could be claimed that he had a more relativistic view of temporal and spatial extension.

However, it was not until Einstein replaced Newton's container spacetime with his relativistic theory, that the observer started to play a role in the description of time. Special relativity allowed for observer frames: what is simultaneous for one observer does not happen at the same time for another. Einstein pointed out the absurdity of a container-like spacetime:

The idea of the independent existence of space and time can be expressed drastically in this way: If matter were to disappear, space and time alone would remain behind (as a kind of stage for physical happening). [4]

Einstein did without an absolute reference system – the constancy of the speed of light acted as a constraint. Simultaneity and duration were defined on the basis of the observer's event horizon and concepts like time dilation only made sense with respect to a specific observer frame. But although the individual perspective determined temporal order, there was no room for the Now in Einstein's model. He differentiated between subjective and objective time and assigned the Now to our consciousness only – to him, it had no physical counterpart. So, although Einstein introduced the observer perspective, this perspective was only defined by an event horizon assigned to a specific position in spacetime. He did, in fact, doubt the reality of the passage of time. Just a few weeks before his own death in 1955, Einstein wrote in a letter to the widow of his friend Michele Besso:

> Now Besso has departed from this strange world a little ahead of me. That means nothing. People like us, believing physicists, know that the distinction between past, present and future is only an illusion, albeit a persistent one. [5]

Einstein's German phrase *für uns gläubige Physiker* has been translated as "people like us, who believe in physics ...". But I prefer my translation because, even though Einstein had, admittedly, never been a follower of the Christian Church and had broken with the Jewish faith, he was still a deeply religious man. He believed not in a God which bothers about the fates of individuals, but in the God of Spinoza, an order and a harmony which shine through all existence. (Einstein's reaction to quantum theory was "God does not throw dice.") It was Einstein's belief that the passage of time, and thus the Now, was a mere illusion. [6]

Enter quantum theory. For the first time, the observer-participant's role in a physical model of the world was more than that of a spectator. By taking a measurement and thus causing the collapse of the wave function, he creates an individual temporal observer perspective, a customized Now, which is realized as a selection of one of many superimposed states. Thus, the measuring process itself is the transition from the possible to the actual. In quantum theory, the observer-participant plays a constituting role. Wheeler points out the selection effect of our questioning Nature:

> Whence the necessity for quantum mechanical observership? Does not the necessity for observership reveal itself in this central feature of nature, that we have no right to say that the electron is at such-and-such a place until we have installed equipment, or done the equivalent, to locate it. What we have thought was there is not there until we ask a question. No question? No answer! [7]

Wheeler formulated the role of the observer-participant as that of a selective questioner. To him, reality did not exist before the observer had posed a question, and therefore had a subjective status. To illustrate the idea of a selecting questioner, we can take a brief look at Wheeler's example of "Twenty Questions", the popular game in which a particular word must be guessed before all twenty questions are up. In Wheeler's version, the questioner asks the first panellist: "Is it edible?". "No," comes the reply. To reduce the number of possibilities, the next question is: "Is it a mineral?" And so on. Then the questioner asks "Is it a cloud?". As usual, there is intense thought among the panellists, but this time the questioner has no idea why no immediate "yes" or "no" answer is forthcoming. Finally, one panellist says "yes" and everybody bursts out laughing. Then they explain to the questioner, that they hadn't agreed on a word beforehand, so each of them had to make sure he could think of a word that fitted all the selections made so far, in addition to his own, before answering "yes" or "no". That is to say, the word was being brought into existence by the chain of questions and answers. It is in the same way, according to Wheeler, that we find out about the electron's position and momentum: the reality we create depends partly on which questions we select to ask.

We used to think that the world exists 'out there' independent of us, we, the observer safely hidden behind a one-foot thick slab of plate glass, not getting involved, only observing. However, we've concluded in the meantime that that isn't the way the world works. In fact, we have to smash the glass, reach in, install a measuring device. But to install that equipment in that place prevents the insertion in the same place at the same time of equipment that would measure the momentum. We are inescapably involved in coming to a conclusion about what we think is already there. [8]

In quantum theory, the Now is an indivisible whole consisting of particles and the experimental set-up, including the measuring chain. While most interpretations of quantum theory proclaim the generation of one world from the successive collapses of the wave function, Hugh Everett saw an alternative, which was fully consistent with quantum formalism. He suggested we interpret also the uncollapsed state vector as objective. This would entail that there are as many simultaneous Nows as there are possible states and interfacial cuts. The price we have to pay by accepting Everett's interpretation is the loss of one world – we would have to recognize the existence of many parallel worlds which exist simultaneously [9]. To some extent, quantum physics takes account of the observer-participant's internal make-up, as he is part of the measuring chain. Apart from Everett's interpretation, however, quantum physics does not explicitly address the status and structure of the Now.

Our understanding of the internal dynamics of a system was changed drastically when Ilya Prigogine defined an internal time T, which differed from astronomical time [10]. This internal time T was an operator, which represented the internal age of a system. In a puff pastry dough, for instance, it would reflect the amount of folding and stretching the system had undergone, rather than some external clock time. T is non-local, as it has to take account of the entire stretching and folding process. Although this was not Prigogine's intention, one may well imagine an observer who consists of the whole puff pastry, embedded in a context with a different internal dynamics, such as someone's stomach. This observer's temporal perspective would have a different structure from that of his embedding context. One effect of such a nesting would be an additional temporal distortion. Prigogine's main point, though, was that we could not associate a local trajectory to the puff pastry, even if we knew the internal age, because T is non-local.

Prigogine's dissipative structures display a property which has become known as sensitive dependence on initial conditions (SDIC). It means that, no matter how precisely we can set up the initial conditions, we cannot determine which way the system will develop. Ed Lorenz noticed SDIC in 1961, when he was re-examining a plotted graph which predicted the weather [11]. The equations for the simulation were simple, involving only three parameters. To re-run the plot, he entered the initial conditions. As computers were very slow at that time, Lorenz left to have a coffee while the system was computing and plotting. When he returned, he was surprised to find that the plot had developed in a manner totally different to the first run. At first, Lorenz believed there had been some malfunction, but after re-running the system, it occurred to him that the reason for the divergent developments was that he had truncated those initial conditions. Instead of typing in .506127, he had shortened it to .506 [12]. This touchiness of dissipative systems, such as the weather, became later known as the "butterfly effect": Does the flap of a butterfly's wing in Brazil set off a tornado in Texas? [13]. The possibility is conceivable, because small changes in such a system are amplified exponentially. Deterministic chaos is defined by the property identified as SDIC. Lorenz pointed out that the concept of initial conditions is a relative one:

"Initial conditions" need not be the ones that existed when a system was created. Often, they are the conditions at the beginning of an experiment or a computation, but they may also be the ones at the beginning of any stretch of time that interests an investigator, so that one person's initial conditions may be another's midstream or final conditions. [14]

The long-term behaviour of dissipative systems can be portrayed in phase space. As briefly mentioned in Chapter 9, many strange attractors display a self-similar fractal structure. However, this self-similarity is a virtual one, as it is only visible in phase space, where each point represents a possible state of the system. What is lost in such a portrayal, however, are the successive vectors which generated the attractor. Δt_{length} is lost along the way and Δt_{depth} is contained only in a virtual nesting, which shows the limitations of the system's long-term behaviour. The fractality of attractors in phase space no longer contains information about the development of events.

Although neither Lorenz nor the pioneers of chaos theory developed a new notion of time or the Now, SDIC remains an important issue, as human beings are dissipative systems too. One direct result is that the way we develop is describable, not by a local trajectory, but only as an ensemble. When we consider an observer-participant's internal complexity in terms of his internal time, we have to take account of as we did with the puff pastry – that is, entirely, and with no short cut. Determining the future behaviour of the rest of the universe has become a matter of resolution. And, although this scale-dependence has emerged for different reasons in SDIC and fractal spacetime, we see in both models different results at different resolutions. We shall return to the topic of scale-dependence after introducing Otto Rössler's interpretation of the Now as pure interface.

It was Rössler who set up the first model which realized that our Now is our only interface with the world. And he introduced the notions of an endo- and an exo-observer (see Chapter 4). He stresses that there is no escape button to release us from our human condition, which imprisons us in space and time – we are confined to our endo-perspective.

Rössler's microconstructivism takes account of intra-observer chaos – that is to say, the internal dynamics of the observer-participant. His internal state will influence the reality he generates. If he is running a temperature or in a threatening situation, he will perceive things differently from the occasions when he is healthy and relaxed. As we have seen, our internal state (which includes conditioning effects, such as making up for delays) and intention even shape the way we perceive duration and temporal order. Rössler even suggests that chaos, as described within a classical framework, may generate a 'fake' quantum mechanics as a result of including the observer-participant's internal chaotic dynamics. His answer to the 'noise in the universe' assumption is the 'noise in the head' proposal [15].

> The quantum world is proposed to represent a special kind of 'interface'. Perspective is an interface, optical illusions are an interface, daily life is an interface, reality is an interface. There is no reality but interface reality. Nowness, too, is pure interface.
> [16]

According to Rössler, the world we see is an interface reality – there is nothing beyond this interface which is directly accessible to us. And this interface is the Now. The entire universe is accessible to us only though our Now – it is all we have.

Newton's description of the universe was undertaken by an exo-observer, who, just like Laplace's demon, was not part of the system he described and had all the information available that was needed to compute the future states of that universe. And although Einstein took account of individual relative perspectives, he did not take account of intra-observer dynamics and therefore did not accommodate a notion equivalent to Rössler's notion of an endo-observer. Furthermore, Einstein did not address the interference inherent in the act of observation. This was taken into account by quantum mechanics, but, again, the dynamics of the observer was not considered in the equation on either side of the Heisenberg cut. Neither did Prigogine's most helpful notion of an internal time include an embedded observer.

The notions of fractal spacetime and scale relativity assign the fractality to spacetime itself (see Chapter 2 and Appendices C-E). However, as the observer-participant selects the scale of observation, he also selects the amount of detail and the number of dimensions he will be able to make out. Selecting the amount of zooming determines whether we look at a system on the classical or the quantum level, as a differentiable or a non-differentiable structure. As Ord has pointed out, unless we are willing to give up differentiability, fractals 'fly under our radar' (see Appendix C). If for instance, an observer looks at El Naschie's e-infinity from a distance – that is to say, at low resolution – he will encounter something that actually looks like a spacetime of only 4 dimensions (see Appendix E). The more he zooms in, however, the more dimensions will reveal themselves to him, until he realizes he is facing an infinite-dimensional fractal spacetime. With the golden ratio as its inherent symmetry, e-infinity spans the net from quantum and relativistic descriptions to the level of our aesthetic perception (as the golden mean is seen as the ultimate expression of harmony and beauty by man, who is himself part of this spacetime – we are stardust). Nottale's scale relativity also models a fractal spacetime whose geometry is resolution-dependent (see Appendix D). He proposes analogies between levels of description such as phylogeny and ontogeny. Cell differentiation, organ formation and tissue growth may all be manifestations of fractal memory which expresses itself on all scales of organization. Together with Timar, Nottale explicitly addresses the constraints of the observer-participant's Now and the relationship between internal and external spacetime limitations [17]. They describe temporal structures from an endo-perspective, acknowledging the scale-relativistic character of our perception of time. Temporal disorders can be described in terms of scale-incompatibility between inside and outside (the endo- and exo-perspective). In my own theory, I have sought to expose the impact of our temporal fractal perspective on our generation of reality.

Rössler's micro-constructivism explicitly addresses the problem of the impact an embedded endo-observer's internal dynamics has on his world-observer interface. Rössler repeatedly stresses that interface reality is all we know and all we can talk about – we have no access to the thing-in-itself and thus always operate on a phenomenological level. So our Now is our very private event horizon. In his interview "Descartes' Dream", Rössler summarizes Roger Shepard's description of this situation: If all Nows are present in my Now, this does not mean that mine is also present in yours [18]. This is an important realization because different endo-perspectives lead to very different spatial and temporal interfacial structures. And as some of us make up for delays and others choose not to or cannot do so, we may even disagree about the order of events: What is cause to observer-participant A may be effect to B, and vice versa. Also the difference between endo- and pseudo-exo-perspectives will impact the way we perceive the world and, for instance, report events (see Fig. 11.1).

Fig. 11.1. An embedded observer-participant (a photographer being photographed). [19]

This paradigm shift, which puts the observer-participant's perspective as an *a priori* of all experience and knowledge, was also proclaimed by German neuroscientist Ernst Pöppel. In Chapter 3, we saw how the nested thresholds of our multimodal sensory inputs form what Pöppel denoted a simultaneity horizon. Pöppel assigns utmost importance to the Now – it is one of our four primary experiences of time, together with succession, simultaneity and duration. He goes further and claims that all physical theories and models of the world we have created are secondary constructs of our primary experiences of time.

> (...) the search for the conditions of rendering possible any experience of time in the real world is determined by mechanisms of our brains, which condition our experience of time. It is not possible to conceive of a theoretical concept of time in physics (e.g. Newton's 'absolute' time), which pretends to exceed experience of time. Therefore, I suggest to regard the physical concepts of time (be it the Newtonian one, that of Einstein or Prigogine) as a secondary construct derived from our primary experience of time. [20]

Pöppel's line of argument is simple: As human beings, we must take into account *a priori* the performances our brain carries out and recognize the fact that all concepts and theories we create are thus necessarily derivatives of our primary experiences of time. He thus concludes that all physical theories are anthropocentric.

The first primary experience of time – simultaneity - we have already encountered in Chapter 3, where we saw how the differing thresholds of our visual, auditory and tactile inputs form a simultaneity horizon. The second primary experience – succession – is also a constraint on our perception. For instance, if events are not separated by at least 30-40 milliseconds, we do not perceive them as successive, but as simultaneous. The third primary experience of time – duration – we have encountered in various manifestations in this book and we have seen how malleable and subjective it is. The last primary experience – the Now – plays a very important role in Pöppel's theory. He claims that the Now is generated by an integration performance carried out by our brains. In other words, we form gestalts, which are indivisible *per definitionem*. Pöppel concludes that the duration of an event depends on the mental capability of the individual who experiences it. The richer and the more differentiated that individual's language, the more complex and extended is his Now. Pöppel's explanation is somewhat reminiscent of Husserl's description of listening to a tune.

> Successive events are perceived as being present only up to a certain limit. An example from language: the word "now" is made up of successive phonetic events. But when I hear the word "now" now, I perceive the whole word now and not a succession of individual phonetic entities. This indicates another performance of brains at work, namely the *integration* of temporally separated events into perception gestalts, which, in each case, are present, i.e. constitute the Now. The upper limit to this integration of perception-related experience ranges between two and four seconds. (...) What we experience as being present is not a point without extension on the time axis of classical physics but meaningful events which have been integrated into gestalts. [21]

Pöppel's concept of the Now is a fractal one – the re-iterated integrations turn the present into an extended, nested structure reminiscent of Husserl's model.

The theories of fractal spacetime and scale-relativity, as well as the notion of an extended observer-participant, share some aspects of David Bohm's notion of an implicate order. In his holistic model, the universe is a hologram – that is to say, an interference pattern of waves. Actually, in a classic hologram, two light beams have to meet to re-create the interference pattern: the object beam and the reference beam. If the observer-participant does not send out a reference beam, no interference pattern will emerge. In such a holographic universe, everything that exists is enfolded within each region of space and time. Thus, no matter how abstract our thoughts, they will still enfold the whole, which can therefore be unfolded in even the tiniest section. Bohm's universe is self-similar, as each part contains the information about the whole. The idea that we can unfold the entire structure of the universe from any level of abstraction is reminiscent of delay representations in phase space, which re-construct the whole attractor in phase space by plotting just one parameter. [22]

Bohm describes how, in contrast to a mechanistic order, the implicate order takes account of the fact that, within a living organism, all parts develop and grow within the context of the whole. Therefore, we may say that no part can exist independently and it does not make sense to try and describe it as an isolated structure or process. In Bohm's model, both matter and consciousness can be understood in terms of the implicate order. He proposes that the basic element of reality be a moment, which spans more or less space and time. Each moment has a particular explicate order and also enfolds all the others. The enfolded forms unfold and thus become manifest to our senses.

> So the relationship of each moment in the whole to all the others is implied by its total content: the way in which it 'holds' all the others enfolded within it. In certain ways this notion is similar to Leibniz's idea of monads, each of which 'mirrors' the whole in its own way, some in great detail and others rather vaguely, The difference is that Leibniz's monads had a permanent existence, whereas our basic elements are only moments and are thus not permanent. [23]

Bohm suggests that mind and body cannot be seen as different entities and are both embedded into a yet higher-dimensional actuality, of which they present mere sub-totalities:

> Each of these [mind and body] is then only a relatively independent sub-totality and it is implied that this relative independence derives from the higher-dimensional ground in which mind and body are ultimately one (rather as we find that the relative independence of the manifest order derives from the ground of the implicate order). [24]

The role of the observer-participant in our models of the universe varies greatly in the different approaches. But there is an unmistakable trend towards putting the steersman back at the helm. The increasing influence of the observer in models of time showed itself first in the replacement of Newtonian absolute time by Einstein's relativistic notion, which allows for observer frames. Then, quantum theory demonstrated that there is no such thing as an independent observer. There are only observer-participants: with every measurement/question, we participate in the generation of reality. Next, Rössler's micro-constructivism introduced the Now as the observer-world interface: we are all endo-observers and our internal make-up and dynamics determines the way we see the world. But as we are trying to define something into which we are embedded – namely, time – we are facing a Gödel limit. We cannot take the necessary step back, we cannot jump out of the system in order to observe it from an exo-perspective. Our embodied minds whose internal processes are of a temporal nature cannot but produce endo-models of the world. Is there a way out? What is required is a Strange Loop in Hofstader's sense. The resolution-dependence of observations in fractal spacetime, as suggested by Ord, Nottale and El Naschie, assigns the observer-participant the power of selecting the scale of observation. Pöppel also makes man the measure of all things when he states that all our physical theories are derivatives of our primary experiences of time and therefore anthropocentric. The same claim has been made, not about physics, but about mathematics, by Lakoff and Núñez, who have shown in numerous examples how mathematical concepts are derived from embodied cognition (see Chapter 4). The resolution-dependence of fractal spacetime and scale relativity places the observer-participant into a decisive position, as he selects the zooming level. The notion of a temporal fractal perspective which can be recognized and possibly altered puts the power to structure time into the hands of the observer-participant. As William Blake put it: "For the eye altering alters all." [25]

In addition, there have been significant developments in cybernetics and systems research. The connection between embodiment and cognition has been widely accepted (e.g. in the development of robots [26]), as well as the fact that we are sentient beings with more or less extended bodies, who live in a physical, social and cultural context.

Consequently, our physical and mental constraints and contexts differ individually, as each person creates his own perspective. This realization is accounted for in 2nd-order cybernetics, which sees the observer-participant as a cybernetic system in its own right. 3rd-order cybernetics explains the observer-participant's subjectivity with the fact that he is nested into a context. The aim of 3rd-order cybernetics is to contextualize the observer-participant ontologies of the 2nd domain. So, bit by bit, the steersman is regaining his influence. His interface with the world has taken centre stage: Enter the Now.

So, why is it always now? The 13th-century philosopher and monk Dogen wrote in great detail about the notion of time and the Now in Buddhism. Michael Eido Luetchford's translations of his texts reveal a highly stimulating philosophy of the Now [27]. To Dogen, the notions of time and existence in the present are equivalent and the present moment, our Now, contains all. It is always now because everything that *is*, is the Now and the Now is all there is. In his translation of Dogen's book *Shobogenzo*, [28]. Luetchford describes how perceiving each thing without evaluating or judging it, allows the experience of a Now which contains all:

> The person I think of as myself is a 'person' that I put together at one time-present. We can apply this thinking to everything in the Universe. This kind of intellectual analysis is the starting point of Buddhist practice. But a person who has clarified their real states sees only each thing, each thing, each thing, and lets go of understanding the nature of each thing. And at that moment, time-present contains the whole of time, and that time contains all things. Thus the whole of existence, the whole Universe, is present at each moment of time. Have a quick look to see if you can find any part of the Universe that has escaped from the present moment. [29]

Buddhist thought believes that only the Now is real. But if the Now contains all, does this mean our past and future are also contained in the Now? What kind of structure can we assign a Now which contains all

[30]? If we equate the Now with everything that exists – that is to say, the universe – then it obviously has the structure of the universe.

Fig. 11.2. Why is it always Now? [31]

The Buddhist has no problem with this explanation at all. Most Westerners, however, would sadly miss their long-term correlations. Surprisingly, East and West can meet after all, if the bridge comes in the shape of a fractal. If we assume a fractal spacetime, as suggested by Ord, Nottale and El Naschie, and equate, in the Buddhist tradition, existence with the Now, then the Now has a fractal structure. But how can something instantaneous have a structure? We usually think of a structure as something extended. As we saw in the last chapter, the Prime has a structure, although it is indivisible – an atom of time, but itself timeless. And the experience of condensation, as in insight, is instantaneous – it has no extension in Δt_{depth}. I may thus venture to define Dogen's Now, in terms of my Theory of Fractal Time, as $\Delta t_{depth} = \infty$ and

$\Delta t_{length} = 0$. It is sheer simultaneity, everything that exists in an instant: the entire universe.

Fig. 11.3. De-nesting at the master level? [32]

The same approach would also hold for some models of fractal spacetime. For instance, Nottale's lower limit, which is the Planck length, is also an extension which can no further be divided – it becomes a universal scale. As Wheeler has pointed out, "at the Planck scale, there is no before and after – the very notion of time loses its meaning." [33]. The different theories propose different "atoms" – that is to say, indivisible modules of spacetime. But as long as those smallest nested structures are extended but indivisible, time and timelessness are connected.

For a beautiful expression of the idea of Now containing the whole of time, we may turn again to William Blake:

> To see a world in a grain of sand
> And a heaven in a wild flower;
> Hold infinity in the palm of your hand
> And eternity in an hour. [34]

But can the idea of an instantaneous reality accommodate, or even assimilate, the idea of a temporally extended observer-participant? The answer may lie in the conclusion that, as long as we generate Δt_{depth} only – that is to say, only simultaneity, rather than succession on one LOD – the content of every Now will be instantaneous. As we have seen, generating Δt_{depth} means contextualizing (or "nesting"). This may happen both by nesting events as retensions or protensions, depending on whether we embed the current event into a remembered past context or an anticipated future one. In Chapter 6, we mentioned Eagleman's claim that all awareness is postdictive. Let us now take a closer look at the connection between anticipation and postdiction.

Although our brains are very good at predicting, this is sometimes not possible in a world which changes unpredictably. This is why, as Eagleman has put it, "brains often decide what happened retrospectively" [35]. He suggests considering postdiction as an alternative or complementary framework to anticipation, as both try to update our picture of the outside world. Both prediction and postdiction follow the goal of compensation, but in different temporal directions. An example of postdiction is apparent motion (e.g. a number of dots lighting up in close succession, which gives us the illusion of one dot moving across a plane). Eagleman stresses that no motion is perceived until the first stimulus has disappeared and the second has appeared in an unexpected place. An example of the opposite type of compensation – that is to say, anticipation – is the key-press experiment described in Chapter 4, in which a removed delay caused a temporal order reversal which was interpreted as a reversal of the causal order. This example of a re-calibration of our motor sensory systems when a delay is inserted or removed shows, in Eagleman's view, that we compensate delays at a very early perceptual levels. The reason why we need this compensation is a direct effect of the way we generate simultaneity:

> The recalibration story begins with a mystery: Given that multi-sensory signals arrive in the brain at different times, how can the brain decide which events were supposed to be simultaneous in the outside world? We have proposed that the brain perceptually

recalibrates its expectations about the arrival times of signals by employing a simple assumption: when the motor system executes an act (such as knocking on a door), all the resulting feedback should be assumed to be simultaneous, and any delays should be perceptually adjusted until simultaneity is perceived. [36]

The idea of an extended Now could thus be defined by a combination of anticipative and postdictive compensation: Contextualization by means of protension or retension extends our Now in the dimension of Δt_{depth}, thus generating a larger number of impressions which we perceive as simultaneous.

Fig. 11.4. The extended Now includes anticipatory compensation by introducing or removing delays. [37]

The notion of an anticipatory system was originally developed by Robert Rosen for living systems and has since been extended to physical systems by Dubois [38, 39]. In particular, the notion of anticipatory regulation has been introduced and widely discussed by Dubois. He has shown that introducing a time shift into a delay system can either generate deterministic chaos or catapult a system out of its chaotic dynamics onto a stable attractor, depending on the size of the time shift. In a system without delay, for instance, introducing an anticipatory time shift can produce chaos, which can then be suppressed by a delay time shift [40]. However, Dubois' notions of incursion and hyperincursion are beyond the scope of this book.

Both regulatory anticipation and regulatory postdiction determine the structure of our Now. A non-balanced distribution of protension and retension can result in pathological conditions, as we have seen in Chapter 7 in the description of the temporal perspectives of depressed individuals. Not being able to re-calibrate one's motor sensory system can also result in serious disorders, and the inability to make up for delays can lead to a perceived reversal of causal order, as in schizophrenia.

Predictive and postdictive compensation are highly useful when it comes to recalibrating rhythms of our Now, as well as initiating or resisting entrainment. Such compensation is a vehicle for restoring physical and mental well-being by successfully embedding an individual's temporal perspective into that of a desired context. This is no pledge to plain conformism but, rather, an invitation to seek out those contexts which allow one to grow and to reject those which suffocate and paralyze.

Bill Seaman has pointed out that each individual has his own structure of nowness [41]. Direct observation is further informed by augmented observation (we have discussed this topic in the context of observer extensions in Chapter 4). Seaman describes the human observer as a nested system which operates and observes on multiple scales:

> The human observer has many interesting characteristics. They represent a physical volume in space that is continuously in motion across multiple scale domains from the micro, to the

macro. Human observers are dynamically nested in an environment that enables them to consciously make observations. They are examples of far from equilibrium dissipative structures, and are thus dynamic systems. They become structurally coupled with their environment to maintain their living status, to make first-hand observations, as well as to communicate about those observations. [42]

Apart from the distortions inherent in the different propagation speed of signals and the varying processing time of stimuli within the body, Seaman also addresses the problem that humans register changes in their environment consciously at a different speed than they do unconsciously. He refers to Benjamin Libet's experiments, which have shown that conscious awareness has a lag of roughly 500 milliseconds [43]. But once we have assimilated the delay, we are no longer aware of it. The only way of making us conscious of it is to remove the delay. As we have seen in the keypress and flash experiment, the removal of an assimilated delay would make us perceive a reversal of the causal order. Libet demonstrated this by directly stimulating parts of the brain, with the result that their owners were convinced they knew about a stimulus before it actually happened.

According to Seaman, every incorporated tool or augmenting machine modifies our understanding of the observed.

Yet we must always remember that we are in this case observing a mediated experience that has attributes unique to its environment, qualities specific to the mode of transduction, and qualities specific to the mode of legibility (visualization and sonification). Again, we ironically decouple the human observer from a time-centric perspective of direct observation in order to more accurately register and parse time via particular tools and machines that take on the observer-centric time frame for us. [44]

Humans can become exo-observers, if they construct and observe a system which hosts both the computational observer and his environment. However, monitoring this double-nesting does not remove the biases which result from our physical and linguistic constraints. Seaman and Rössler have introduced the notion of "neosentient machines" – machines which are not only embodied and embedded but can also compensate for delays and, above all, have mirror competence. This means they are capable of simulating their environment – that is to say, they are empathic. The device is called "The Benevolence Engine." It functions as a measuring system, a computational simulation system, a time-parsing system (at scales beyond human perception) and has introspective capabilities. The idea is that the machine would be a first-hand observer to measurements which span scales inaccessible to direct human observation. In contrast to a human observer, such a neosentient machine would not need to be de-coupled from the event it observes.

Machines can help to link us to time scales beyond our reach. Sometimes, their only *raison d'être* is to make us aware of time spans we would normally not consider. These intervals may be nanoscale – such as the time it takes for a photon to cross a living room (about 10 ns) – or they may cover aeons, as does a geological change like a continental drift. An example of a device whose sole purpose is to make us aware of time scales beyond our reach is the prototype of the Millennium clock. This is a project initiated by The Long Now Foundation, established in 1996 to promote "slower/better" thinking as opposed to today's "faster/cheaper" mindset and to foster responsibility for the next 10,000 years. [45]

> Civilization is revving itself into a pathologically short attention span. The trend might be coming from the acceleration of technology, the short-horizon perspective of market-driven economics, the next-election perspective of democracies, or the distractions of personal multi-tasking. All are on the increase. Some sort of balancing corrective to the short-sightedness is needed – some mechanism or myth which encourages the long view and the taking of long-term responsibility, where 'long-term' is measured at least in centuries. Long Now proposes both a mechanism and a myth. [46]

The Millennium clock is just one of the Long Now Foundation projects: it is a large mechanical clock, which is powered by seasonal temperature changes. This clock is conceived as a high-precision device which will deviate only 1 day in 20,000 years and self-corrects by means of entrainment: it phase-locks into the noon sun. The Millennium Clock "ticks once a year, bongs once a century, and the cuckoo comes out every millennium." [47]. The Foundation has bought part of a mountain eastern Nevada, where the final gigantic structure will be situated. In the meantime, a small-scale prototype is on display at the London Science Museum and a second is under construction. The idea is to make visible a framework and perspective people are usually not aware of: "Ideally, it would do for thinking about time what the photographs of Earth from space have done for thinking about the environment. Such icons reframe the way people think." [48]

Raising awareness is the first step to changing a mindset. The Long Now Foundation attempts to embed people's fleeting Now into a more extended one – in fact, into a Now which spans 10,000 years and whose beginning and end is marked by the appearance of a cuckoo. We have pointed out the advantages of nesting, but de-nesting also has its uses. Witness the de-nesting exercise of meditation, which can be seen as a counterweight to the Long Now Foundation's nesting project. Meditation allows us to rid ourselves of different layers of reality, of mere appearances and attributes, of thinking and perceiving. The eight stages of *dhyana*, which may be translated as concentrating or focussing, are an example of a step-by-step de-contextualization, detachment ("de-nesting") [49]. But it is not only through the activities of nesting and de-nesting that we determine our physical and mental well-being. Entrainment of internal and external rhythms is also a powerful influence on our condition. Some rhythms we can seek, generate, evade or modify. Others are beyond our sphere of influence. We are now beginning to regard many illnesses as dynamical diseases and have recognized the importance of a rich range of internal rhythms. And we are capable – at least to some extent – of reflecting our human condition. But we are still a far cry from anticipating the full extent of the structurability of time. To get a hunch of what we may be missing, let us reflect on the words Billy

Pilgrim's favourite science fiction author wrote about Earthlings in his book called "Maniacs in the Fourth Dimension":

> It was about people whose mental diseases couldn't be treated because the causes of the diseases were all in the fourth dimension, and three-dimensional Earthling doctors couldn't see those causes at all, or even imagine them. [50]

References

1. Kurt Vonnegut, *Slaughterhouse-Five*, Dell Publishing, New York 1991, p. 115 (first published in 1966).
2. J.A. Wheeler, Time Today, in: J.J. Halliwell, J. Pérez-Mercader, W.H. Zurek (Eds.), *Physical Origins of Time Asymmetry*. Cambridge University Press 1994, p. 8.
3. I. Newton, *Mathematical Principles of Natural Philosophy*, translated by A. Motte, revised and annotated by F. Cajori. University of California Press, Berkeley 1962, Vol. 1, p. 6 (first published in 1687).
4. A. Einstein, *Relativity – The Special and the General Theory*, Methuen, London 1988, p. 144 (first published in 1920).
5. H.-J. Störig, Die Zeit – eine Illusion?, in: *Scheidewege*, Band 36, Jahrgang 2006/2007.
6. ibid.
7. J.A. Wheeler, Time Today, in: J.J. Halliwell, J. Pérez-Mercader, W.H. Zurek (Eds.), *Physical Origins of Time Asymmetry*. Cambridge University Press 1994, p. 19.
8. ibid.
9. H. Everett, 'Relative State' Formulation of Quantum Mechanics, in: *Rev. Mod. Phys.* (1957), Vol. 29, pp. 454-462.
10. I. Prigogine, *Vom Sein zum Werden*. Piper, Munich 1985.
11. E. Lorenz, Deterministic Nonperiodic Flow, in: *Journal of the Atmospheric Sciences*, Vol. 20 (March 1963) pp. 130-141.
12. J. Gleick, *Chaos: Making a New Science*, Heinemann, London 1988, pp. 16-18.
13. E. Lorenz, *The Essence of Chaos*, University of Washington Press 1993, pp. 181 ff.
14. ibid, p. 9.
15. O.E. Rössler, Intra-Observer Chaos: Hidden Root of Quantum Mechanics?, in: M.S. El Naschie, O.E. Rössler and I. Prigogine (Eds.), *Quantum Mechanics, Diffusion and Chaotic Fractals*, Pergamon Press 1995, p. 105.
16. ibid, p. 107.
17. L. Nottale and P. Timar, Relativity of Scales: Application to an Endo-Perspective of Temporal Structures, in: S. Vrobel, O.E. Rössler, T. Marks-Tarlow (Eds.),

Simultaneity – Temporal Structures and Observer Perspectives, World Scientific, Singapore 2008.

18. O.E. Rössler, *Descartes' Traum. Von der unendlichen Macht des Außenstehens.* Audio-CD. Supposé, Cologne, 2002.

19. Photograph courtesy of Ian, 2010. Original title: Caught in Action. Flickr stream: M.A.M08.

20. E. Pöppel, Erlebte Zeit und Zeit überhaupt: Ein Versuch der Integration, in: H. Gumin, and H. Meier (Eds.), *Die Zeit. Dauer und Augenblick.* Piper, Munich 1989, p. 380, my translation.

21. ibid, p. 372, my translation.

22. H. Kantz and T. Schreibner, *Nonlinear Time Series Analysis*, Cambridge University Press, Cambridge, UK 1997, pp. 127-134.

23. D. Bohm, *Wholeness and the Implicate Order*, Routledge, London 1980, p. 261.

24. ibid, p. 265.

25. J. Sampson, *The Lyrical Poems of William Blake*, Oxford at the Clarendon Press 1906, p. 129.

26. R. Pfeiffer and J. Bongard, *How the Body Shapes the Way we Think.* MIT Press, London 2007.

27. M.E. Luetchford, The Concept of Now in Dogen's Philosophy, in: S. Vrobel, O.E. Rössler, T. Marks-Tarlow (Eds.), *Simultaneity – Temporal Structures and Observer Perspectives*, World Scientific, Singapore 2008.

28. *Shobogenzo* Chapter 11 – Uji. Translation by M.E. Luetchford 2004, www.dogensangha.org.uk.

29. ibid, p. 2.

30. M.E. Luetchford, The Concept of Now in Dogen's Philosophy, in: S. Vrobel, O.E. Rössler, T. Marks-Tarlow (Eds.), *Simultaneity – Temporal Structures and Observer Perspectives*, World Scientific, Singapore 2008, p. 30.

31. Illustration courtesy of Tom Davies, 2009.

32. Photograph courtesy of Stephanie Young Mertzel, 2010 (Original title: "Buddha Dubois & Dogen Quote").

33. J.A. Wheeler, Time Today, in: J.J. Halliwell, J. Pérez-Mercader, W.H. Zurek (Eds.), *Physical Origins of Time Asymmetry.* Cambridge University Press 1994, p. 11.

34. W. Blake, Auguries of Innocence, Nicholson & Lee (Eds.) The Oxford Book of English Mystical Verse, 1917 (first published in 1863).

35. D.M. Eagleman, Prediction and Postdiction: Two Frameworks with the Goal of Delay Compensation. Commentary to: R. Nijhawan, Visual Prediction: Psychophysics and Neurophysiology of Compensation for Time Delays, in: *Behavioural and Brain Sciences* (2008), Vol. 31, p. 205.

36. ibid, pp. 205-206.

37. Photograph courtesy of Rob Oechsle Collection, 2010.

38. R. Rosen, *Anticipatory Systems: Philosophical, Mathematical and Methodological Foundations.* Oxford, Pergamon Press (1985).

39. D.M. Dubois, Introduction to Computing Anticipatory Systems, *International Journal of Computing Anticipatory Systems*, CHAOS, Liège, Belgium, Vol. 2 (1998), pp. 3-14.
40. D.M. Dubois, personal communication 2006.
41. B. Seaman, Unpacking Simultaneity for Differing Observer Perspectives, in: S. Vrobel, O.E. Rössler, T. Marks-Tarlow (Eds.), *Simultaneity – Temporal Structures and Observer Perspectives*, World Scientific, Singapore 2008.
42. ibid, pp. 244-245.
43. B. Libet, *Mind Time*. Harvard University Press, London 2004.
44. B. Seaman, Unpacking Simultaneity for Differing Observer Perspectives, in: S. Vrobel, O.E. Rössler, T. Marks-Tarlow (Eds.), *Simultaneity – Temporal Structures and Observer Perspectives*, World Scientific, Singapore 2008, p. 252.
45. http://www.longnow.org
46. ibid.
47. ibid.
48. ibid.
49. D.R. Komito, *Nagarjuna's "Seventy Stanzas". A Buddhist Psychology of Emptiness.* Snow Lion Publications, Ithaka, 1987.
50. Kurt Vonnegut, *Slaughterhouse-Five*, Dell Publishing, New York 1991, p. 104 (first published in 1966).

Chapter 12

Outlook: Here Be Dragons

When the old seafarers encountered uncharted territory, they would find those blank areas on the map marked with the phrase "Here be Dragons" and the image of a sea serpent or a similarly ferocious creature. Modern steersmen have different images and metaphors for the unknown, such as undefined areas beyond event horizons or singularities. There are some who prefer religious explanations to scientific ones and others who manage to reconcile the two. What they all have in common is a strategy of 'angst reduction': by naming or picturing the unknown, it becomes graspable, albeit often as a mere negation of the known [1]. This is a necessary step, as we can only define a boundary if we know that something exists beyond it, however intangible that something may be.

In the course of history, few people have known much about sea serpents. But an astonishing amount of detail has been known about a close relative of such creatures and the most dangerous beast ever: the mythical basilisk, king of the serpents. It has been described for 2000 years and, in the Middle Ages, appeared in several European cities, including Warsaw, Vienna and Basel. As everybody knew, it emerged from a small, abnormal egg laid by an old rooster and hatched by a toad. It was highly poisonous and could kill with a single glance, so that eye contact had to be avoided at any cost. Reputedly, its glance would also petrify – that is to say, it would freeze its victim in time. The basilisk is the heraldic emblem of Basel (see Fig. 12.1), possibly because of the similarity of the names: "Balsilea", the old name for the city, and "basilisk" mean "the royal" and "little king," respectively [2]. It was there that, in 1474, an aged rooster was seen laying an egg, whereupon it was captured and tried. After being convicted of an unnatural act, it was burned alive before thousands of citizens who watched its execution [3].

The criminal prosecution of animals was not a rare event in the Middle Ages. To the Catholic Church, the basilisk was the manifestation of evil, even the devil incarnate. The Church made the basilisk responsible for such disparate events as the devastating earthquake of 1356 and the demise of the Bishop of Basel during the reformation [4]. Such shared views (as nurtured by the Church) and direct clerical reactions were then a valid form of angst reduction. This framework enabled people to reduce the complexity of the unknown and to anticipate each other's behaviour. And if we smile at this or shake our heads in disbelief, we should keep in mind that in a thousand years, people will be likely to react similarly when they learn about our urban legends, the research projects of CERN or the various endeavours of the Catholic Church.

In order to examine anticipation as an angst-reducing strategy and our ability to come to terms with complexity, let us look at how people used to defend themselves against the basilisk.

Fig. 12.1. The Basel Basilisk. [5]

Though deadly, the basilisk was not invincible, for it had only to gaze at its own reflection in a mirror and it would die instantly. It is said that Alexander the Great used a mirror against a basilisk that was defending a city and Saint George used his shield to reflect another basilisk's image [6]. And in the famous legend of the Warsaw basilisk, a man wearing an outfit of mirrors finally defeated it. It is not known how the basilisk's gaze dealt death to its victim. So a little speculation may be allowed. An intelligent guess, based on the assumption that we simulate our environment, would be that the reptile's fixed stare caused its victim's mirror neurons to perform the same action, namely, to return the same unblinking gaze, in a petrified gaze. In this hypnotic state, the victim would perceive nothing but the gaze of the basilisk – his temporal perspective would be reduced to one LOD.

If the victim could withstand the gaze for a few seconds, the basilisk would also have to endure the other's stare. And, as we know that to look into a mirror would kill the basilisk instantly, a human who simulated the creature's gaze may have been able to achieve the same result. The only problem would have been how to bridge the first few seconds. A possible solution can be found in an idea originating in *A Different Kind of Darkness*, a science fiction story by David Langford. In it, a natural immunization is reached through conditioning exercises.

Although we are all familiar with the adage that looks can kill, the idea of a visual pattern causing instant death is fairly new. Langford's 2004 story contains the best known of such patterns. This, the so-called "Parrot," is neither a picture of a parrot nor that of a basilisk, but a fractal pattern, which is so complex that it blows the brains of everyone who looks at it [7]. The idea is that this pattern breaks through Gödel limits and contains Strange Loops, which the brain cannot process. Logical contradictions and circularity in language can be filtered out. But once the visual pattern has hit the retina, there is no time to drop an iron curtain. Such a BLIT weapon (BLIT stands for "Berriman Logical Imaging Technique") catches the brain unprepared and does irreparable damage instantly. Interestingly, it turns out that exposure to such deadly patterns can be trained and that, consequently, immunity can be attained. The Shudder Club, a gang of kids in the story, did just that: they forced themselves to look – repeatedly and for as long as possible, without collapsing – at a BLIT pattern which was less complex than the feared "Parrot." By this tactic, they became able to withstand the pattern for a few seconds. In situations where grown-ups would collapse instantly after just a glance at the pattern in question, two of the gang members could withstand it long enough to ... But no, I won't spoil the story for you.

One cause of what might have happened during the gang's conditioning exercises was a gradual de-activation of their mirror neurons. (The gang's slogan was "That which does not kill us makes us stronger".) In the story, this ability to de-activate turned out to be a life-saving skill. We saw in Chapter 8 how doctors can protect themselves from simulating their patients' pain by switching off or driving down their own mirror neuron activity. This can lead to a reduction in empathy, which patients may feel and react to, as Gladwell describes. It may well tip a patient's decision to sue a doctor who does not show compassion or it might prevent a college student from appreciating empathy, as we have seen in Chapters 8 and 9. For some occupations, such reduction of mirror neuron activity is not an idle, arbitrary choice, but a necessary (if unconscious) self-protective measure. However, mirroring *per se* is not necessarily positive. To simulate the wrong environment and behaviour can be damaging, as Friedrich Nietzsche knew:

He who fights with monsters should be careful lest he thereby become a monster. And if thou gaze long into an abyss, the abyss will also gaze into thee. [8]

But while we may reflect on the possibility of defeating the deadly complexity of the fractal BLIT images of the story by slowly conditioning ourselves to that level of complexity, there is no known immunity if the gaze of the basilisk hits you unprepared. If our range of responses is too narrow, there seems to be only one line of defense: to deactivate our mirror neurons altogether. This is a real – if highly questionable – option, as individuals lose, together with their empathic skills, their ability to simulate their environment. A direct result will be that these individuals' ability to anticipate behaviour in their immediate vicinity becomes highly compromised.

What happens when we cease to simulate our environment? Michael Zak has developed a mathematical model for simulating such a state [9].

He assumes that human behaviour is governed by feedback from the external world. When we lack external cues, however, we rely on what he calls "common sense".

> (...) when the external world does not provide sufficient information, a human being turns for 'advice' to his experience, and that is associated with common sense. The simplest representation of human experience is if-then rules. However, in the real world, the number of rules grows exponentially with the dimensionality of the external factors, and implementation of this strategy is hardly realistic. One of the ways to fight such a combinatorial explosion is to represent rules in a more abstract form by removing insignificant detail and making the rules more inclusive. (...) Indeed, many natural and social phenomena exhibit some degree of regularity only on a higher level of abstraction, i.e. in terms of some invariants. (...) this means that only variables of the highest level of abstraction are capable of classifying changes in the external world. [10]

Zak understands common sense as feedback from experience. His model is a phenomenological one, representing "the ability to communicate via information flows, and the ability to increase their complexity (...) without interactions with the external world" [11]. The resulting decrease in complexity is entirely self-made as people re-iterate their own thoughts for want of externally provided information.

Is it possible, as Zak suggests, to simulate the world on some scales but not on others? It is, indeed: we do it all the time, when forming gestalts and using metaphors, and thereby reducing the complexity of the world. But normally, of course, we can also rely on the influx of information from the external world. Relying too much on memory and not enough on anticipation forms a retention-heavy Now, which, as we have seen, is a hallmark of depression. It is not only a question of the level of abstraction we select for our models of the world, but also a matter of the distribution of protensions and retensions, which determines the healthiness of our temporal perspective.

So if we have no time to adapt and find ourselves mirroring an evil eye or a scenario too gruesome to take, we may de-activate our mirror neuron activity altogether, in order to survive. This, however, comes at the cost of being unable to correctly anticipate behaviour in our environment. If we de-activated our mirror neurons and ceased to simulate the world, no protensions would be added to our Now, due to our inability to anticipate. The argument for this proceeds as follows:

(1) There is no anticipation without simulation of the external world via mirror neurons.
(2) If no new rhythms are acquired, our range of responses, our internal complexity, is diminished.
(3) Internal re-iteration of experience involves only one's own dynamics: only retensions are nested.
(4) Without anticipation, no protensions are formed in the Now.
(5) Therefore: This retention-heavy distribution leads to a pathological perspective, as one is left to boil in one's own broth.

Provided our empathic abilities are not compromised, we automatically simulate our environment. But mere simulation does not mean incorporation. As we saw in Chapter 4, more is needed to form a new systemic whole – namely, a re-mapping of the brain. Here, Piaget provides useful categories. He proposes that we rely on

> intelligent perception in terms of two complementary movements, *accommodation* and *assimilation*. From the roots 'mod', meaning 'measure', and com', meaning 'together', one sees that to accommodate means 'to establish a common measure' (...). Examples of accommodation are fitting, cutting to a pattern, adapting, imitating, conforming to rules, etc. On the other hand, 'to assimilate' is 'to digest' or to make into a comprehensive and inseparable whole (which includes oneself). Thus, to assimilate means 'to understand'. It is clear that in intelligent perception, primary emphasis has in general to be given to assimilation, while

accommodation tends to play a relatively secondary role in the sense that its main significance is as an aid to assimilation. [12]

Our mirror neurons create Δt_{depth} in the form of nested protensions. If they are dysfunctional, we can no longer anticipate the behaviour of our environment. No simulation: no simultaneity. When our mirror neurons are adequately functioning, both accommodation and assimilation occur (usually in that order).

As shown in Table 12.1, accommodation leads to an increase in Δt_{depth} when we contextualize our perceptions within the framework of a non-assimilated and, therefore, external, dogma or paradigm. With every new nesting, the structure of our Now becomes more complex. However, accommodation points to an initial lack of internal complexity within the observer-participant: he has only a narrow range of responses on few scales. His observer extensions are merely add-ons: no boundary shifts occur, therefore, no new systemic wholes emerge and internal boundaries are visible. There is no compensation for delays – that is to say, there is no endo-anticipation. Observer extensions are not yet incorporated (although they may become so in due course, when assimilation sets in). Examples of this can be found in a variety of forms ranging from a new sports car or pair of spectacles, to a faster broadband connection or the first tentative steps towards trust. Accommodation as a mode of experience will not suffice to withstand the gaze of the basilisk, as its complexity cannot be reduced by forming gestalts or making up for delays.

By contrast, assimilation leads to an decrease in Δt_{depth}: we form gestalts and metaphors and thereby make the world less complex. This becomes possible as a result of our high degree of internal complexity – that is to say, our wide range of responses on a large number of scales. In the process of true incorporation, boundary shifts occur and new systemic wholes are formed. Internal boundaries become transparent and delays are compensated (endo-anticipation occurs). Observer extensions are truly incorporated – that is to say, the body map is rewritten. Examples of this rewriting include the assimilation of tools (such as the stick which makes a hidden hemisphere accessible), compensated delays (like the timely interception of a trajectory) and the achievement of trust.

Table 12.1. Manifestations of accommodation and assimilation.

	Accommodation	Assimilation
Characteristics	Dogma/paradigm provides context, no new systemic whole, no boundary shift	Context is incorporated ("digested"), formation of new systemic wholes, boundary shifts
Type of nesting, structure of our Now	Complexity-increasing nesting: $+ \Delta t_{depth}$, Lack of internal complexity (a narrow range of responses on few scales)	Complexity-reducing nesting: $- \Delta t_{depth}$ A high degree of internal complexity (a wide range of responses on a large number of scales)
Visibility of boundaries and delays	Visible boundaries, delays are not compensated and noticable	Transparent internal boundaries, delays are compensated, become transparent
Type of anticipation	Dubois' notions of exo-anticipation	Dubois' notion endo-anticipation
Observer extensions	Add-on observer extensions	Incorporated observer extensions, re-written body map
Examples	New spectacles, faster broadband connection, first steps of trusting no ability to withstand the gaze of the basilisk, no control over duration	Assimilated tools, compensated delays, trust ability to return the gaze of the basilisk, ability to modify duration

Assimilation may be sufficient to return the gaze of the basilisk, as its complexity can be reduced by forming gestalts and anticipating delays. We saw this at work in Malcolm Gladwell's *Blink!*, where thin-slicing became possible for trained and experienced individuals (see Chapter 10).

So what is the difference between returning the basilisk's gaze and watching a heating kettle? The brain-blasting complexity of the BLIT images or the basilisk's gaze would require a degree of Δt_{depth} humans are almost incapable of acquiring – even with training, it is only possible to withstand the gaze for a while (as the members of the Shudder Club did). The heating kettle, on the other hand, is an image of one-dimensional focussing, of shuffling ahead on one LOD, thus producing Δt_{length}.

Surprisingly, a high degree of internal complexity – that is to say, the observer-participant's range of responses on different scales (LODs) – is the antidote to both meeting the basilisk's gaze and watching a heating kettle. When we are overwhelmed by external complexity (as if facing a BLIT image), a fractal temporal perspective allows us to form gestalts and thereby diminish Δt_{depth} and, in doing so, reduce the complexity of the world around us. By contrast, if we are bored stiff because we are stuck on one LOD (as when watching and waiting for a kettle to boil), the fractal structure of our Now allows us to contextualize – that is to say, to embed one action into another and thereby to create Δt_{depth} and reduce Δt_{length}. This way, gestalts are formed, complexity is reduced and Δt_{depth} diminished so that a complex scene becomes manageable. A scene becomes digestible when contextualizations create nestings of LODs, which increase Δt_{depth} in order to reduce Δt_{length}.

And, although BLIT images are, up to now, part of the fabric of science fiction, we can all imagine incidents whose sheer complexity and speed leave us powerless, almost paralyzed, as mere bystanders forced to concede that control is no longer an option. This is as true, for instance, of the dynamics in financial markets as it is for any operating system whose development and employment are too complex for any one individual to have a comprehensive overview. Admittedly, in comparison with such examples, the problem of being bored by a kettle in the kitchen seems to be a small one. However, when carried to its extreme – say, an individual suffering from sensory deprivation – our body map may change and transform the individual into a kind of vegetable, no longer capable of catapulting himself out of his narrowed existence. Then Δt_{length} becomes sheer torture. To a lesser degree, we all frequently encounter this phenomenon, when we find ourselves bored to tears or trapped in an unvarying environment such as the same company from day to day. ("L'enfèr, c'est les autres" [13]). However, we know that if our ability to simulate and anticipate is compromised, a retension-heavy Now will severely limit our perspective. In this case, deep nesting (a high degree of internal fractality) through active contextualization – e.g. socializing – is the antidote.

A small step towards structurability of time has been presented. Our temporal perspectives are structured by the distribution of Δt_{depth} and Δt_{length} in our Now. It is up to us to re-distribute and re-calibrate.

I would propose that duration arises from friction, which is brought about by too little Δt_{depth}. We can lubricate a system or grind it to a halt by adding or removing Δt_{depth} [14].

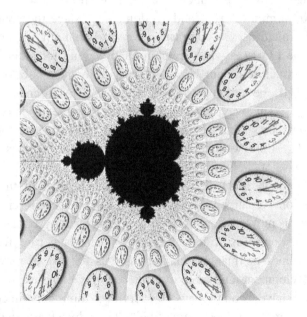

Fig. 12.2. Duration is friction resulting from a lack of contextualization (a lack of Δt_{depth}). It can be lubricated by 'fractalizing' our Now. [15]

As we have seen, trust is both our most powerful means of anticipation and a lubricator, as it reduces Δt_{depth}. This is good news, as it is beneficial not only for the individual who brings it forward, but for everyone involved – a win-win situation.

To conclude, duration is a kind of friction we can modify by distributing and re-distributing the dimensions of Δt_{depth} and Δt_{length} according to our needs. This may be done by shifting our boundary outward or inward – that is to say, by assimilating or removing tools or by compensating or removing delays. Generating duration is a matter of nesting and de-nesting: seeking out or rejecting contextualizations, in particular, in the form of compensating or removing delays. Or, in the apt words of my friend Otto van Nieuwenhuijze: Reality is a realization.

References

1. It was Otto van Nieuwenhuijze who drew my attention to the concept of "anxiety reduction", personal correspondence 2008. accessed 8 July 2010.
2. Bedeutung des Basilisk, in: *Fragen zum alten Basel.* http://www.altbasel.ch/fragen/basilisk.html, accessed 8 July 2010.
3. The Strange Tale of the Warsaw Basilisk. CFI Blogs. http://blogs.forteana.org/node/107, accessed 8 July 2010.
4. *New World Encyclopedia*: Basilisk, http://www.newworldencyclopedia.org/entry/Basilisk, accessed 8 July 2010.
5. Photograph courtesy of Wolfram Krieger 2010. (Flickr: Wolfsraum)
6. *New World Encyclopedia*: Basilisk, http://www.newworldencyclopedia.org/entry/Basilisk, accessed 8 July 2010.
7. D. Langford, *Different Kinds of Darkness*, Cosmos Books, Holicong, PA, 2004.
8. F. Nietzsche, *Beyond Good and Evil*. Apophthegms and Interludes (Chapter IV) 146. Digireads.com, Stilwell, KS, 2005.
9. M. Zak, Modelling Common-Sense Decisions, in: S. Vrobel, O.E. Rössler, T. Marks-Tarlow (Eds.), *Simultaneity – Temporal Structures and Observer Perspectives*, World Scientific, Singapore 2008.
10. ibid, p. 302.
11. ibid, p. 305.
12. D. Bohm, *Wholeness and the Implicate Order*, Routledge, New York 1980, p. 178.
13. J.-P. Sartre, *Huis Clos*. Gallimard, Paris 1947.
14. S. Vrobel, *The Joys and Woes of De-Nesting*. Talk held at InterSymp 2010, Baden-Baden, Germany.
15. Illustration courtesy of Mimi (Flickr: Eskimimi) 2010.

Fractal Dimensions

A fractal is defined as a structure whose Hausdorff Dimension exceeds its topological dimension [1]. What is the difference between these two dimensions? The topological dimension of a point is 0, a line has a topological dimension of 2 and a cube is 3-dimensional. The Hausdorff Dimension for, say, a structure in 3-dimensional space is obtained by covering that structure with spheres (similar to the Box-Counting Method). The smaller the radius (R) of those spheres, the larger the number (N) needed to cover the structure. When R approaches 0 – that is to say, the extension of a point – the Hausdorff Dimension may be calculated as $D = -\lim R \to 0 \, (\log (N) / \log (R))$.

Let us look at a couple of examples. A differentiable curve, say, a straight line, whose topological dimension is 1, would also have a Hausdorff Dimension of 1. A non-differentiable one like the Koch curve, however, has a Hausdorff Dimension of 1.2619... . Therefore the Koch curve is a fractal. For a cube with a topological dimension of 2, the Hausdorff dimension is also 2, therefore it is not a fractal. A broccoli, however qualifies as a fractal, because its Hausdorff dimension (2.66) exceeds its topological dimension.

For self-similar structures, the Hausdorff Dimension $D = \log(n)/\log(s)$. This is the similarity dimension we calculated for the Koch curve in Chapter 2. It is easy to detect whether a structure is a fractal or not, because only fractals have a broken dimension. If we look at a square, which consists of 9 smaller squares of 1/3 of its size, we can see that $D = \log 9/\log 3 = 2$ (It is also possible to use any other logarithm). The Koch curve, by contrast, has a similarity dimension $D = \log(4)/\log(3) = 1.2618$. (One may use circles or squares for 2-dimensional structures and spheres and cubes to cover 3-dimensional ones.)

For fractals whose generating rule cannot easily be made out, the Box-Counting dimension is the best approach. Michael Barnsley, who conceived this method, stressed that, in his space, even a cube of dimension 2 could be defined as a fractal, because it was rather the method than the shape of the structure to be measured, which defines a fractal in Barnsley's approach. One may say, that we put on fractal spectacles when we follow the Box-Counting Method, and thus create a fractal spacetime.

Barnsley defines a Euclidean metric space, in which a structure is covered by closed just-touching boxes of side length $(1/2^n)$. N_n (A) denotes the number of boxes of side length $(1/2^n)$ which intersect the attractor (that is to say, the structure in this space which is covered by boxes). If D = lim n $\rightarrow \infty$ {ln $(N_n$ (A)) / ln (2^n)}, then A has fractal dimension D [2].

In Barnsley's metric space, any structure can be described in terms of its scaling properties, so even integer dimensions are seen as fractals. Plane-filling curves also have a Hausdorff dimension of 2 – that is to say, not a broken dimension – and still qualify as fractals. The Peano and the Moore curve are examples of such structures. These are, so to speak, not only mathematical 'monsters', as Brownian motion is also plane-filling and has a Hausdorff dimension of 2. Such structures may be used to define a fractal spacetime.

But let us take a look at some more familiar structures and their Hausdorff dimension [3].

Structure	Fractal dimension
Coastline of Great Britain	1.25
Coastline of Norway	1.52
Distribution of galaxy clusters	~ 2
Cauliflower	2.33
Broccoli	2.66
Surface of human brain	2.79
Lung surface	2.97

References

1. B.B. Mandelbrot, The Fractal Geometry of Nature, W.H. Freeman, San Francisco 1982.
2. M. Barnsley, Fractals Everywhere, Academic Press, London 1988, pp. 176-177.
3. Wikipedia: *List of fractals by Hausdorff dimension*, accessed 17.07.2010. http://en.wikipedia.org/wiki/List_of_fractals_by_Hausdorff_dimension

Appendix B

Using the Box-Counting Method to Determine the Fractal Dimension of the Koch Curve

In Chapter 2, I have used the self-similarity dimension to calculate the fractal dimension of the Koch curve. This was easy, as I knew both the initiator and the generator. But if I do not know the number of parts per LOD or the scaling factor, the Box Counting Method is a good alternative. However, as we saw in Chapter 2, when we tried to approach the fractal dimension of the Labrador coast, this method is messy, and one needs to refine the boxes many a time in order to reach an acceptable approximation.

We have calculated the fractal dimension of the Koch curve in Chapter 2 by means of the self-similarity dimension:

$$d = \ln \text{ (number of parts) } / \ln \text{ (scaling factor)}$$
$$= \ln 4 / \ln 3$$
$$= 1.2618 \dots .$$

Below, are the first 4 steps of the Box Counting Method. We count only the boxes which are intercepted by the line of the Koch curve, take the logarithm of that number and divide it by the logarithm of the scaling factor.

For squares of side length 1/6: ln 14 / ln 6 = 1.4738 ...

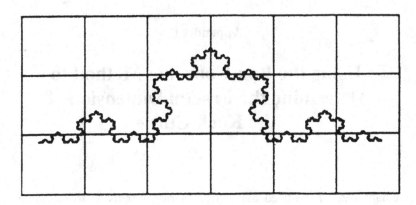

For squares of side length 1/12: ln 26 / ln 12 = 1.3111 ...

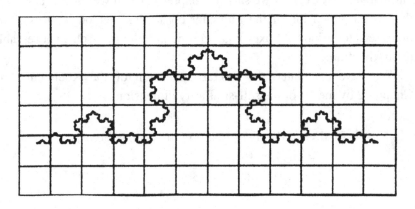

For squares of side length 1/24: ln 58 / ln 4 = 1.2776 ...

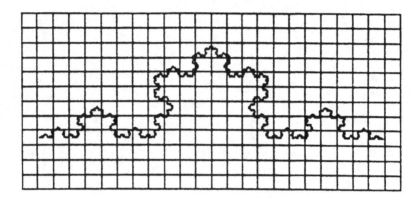

For squares of side length 1/48: ln 134 / ln 48 = 1.2651 ...

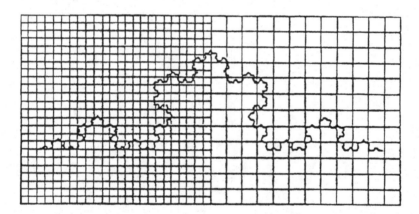

The last zoom gets close to the value determined by means of the self-similarity dimension (1.2618 ...)

Appendix C

Even After 25 Years, Fractal Spacetime is Still Odd

G. N. Ord

The Fractal Spacetime paper of 1983 [1] (henceforth FST) was an attempt at reconciling a particle-path 'picture' of quantum mechanics with the Heisenberg uncertainty principle. Although this had been done with great skill and insight in the 1940s by Richard Feynman in his path-integral formulation, the result had been so successful as a calculational scheme that its potential for providing an explanation for wave-particle duality had been somewhat neglected. In the intervening time, Fractals had made an appearance [2] and FST showed that the scaling relations involving wavelengths in quantum mechanics had a natural reinterpretation in terms of Fractals. These elements of scale have been explored in more depth, particularly in the works of Laurent Nottale and Mohamed El Naschie.

Two general features became apparent from the use of Fractals. The first was that if one considered particle paths in non-relativistic quantum mechanics, the Fractal dimension of those paths would be 2. This suggested that quantum propagation shared a similarity with diffusion since the Fractal dimension of Wiener paths was also 2. Although this was not a surprise, given the similarity between the Schrödinger equation and the diffusion equation, it did serve to suggest that the uncertainty principle was *not* strictly a harbinger of quantum mechanics, it was an indication of underlying Fractal geometry, shared with many classical systems.

A second feature of FST was that the central concept distinguishing quantum mechanics from classical mechanics was the need to couple both space and time through Fractals. Fractal space combined with time as a simple parameter would simply not support interference. Part of the significance of FST was that if one were to look for a genuine underlying statistical mechanics for quantum mechanics, that search would involve a form of Fractal time. However, in a differential formulation, Fractal time itself could ultimately evade detection underneath a continuum limit. Just as Fractal paths underlie, but are not detected by the differential operators in either the diffusion or Schrödinger equations, a form of Fractal time appears to underlie both the Schrödinger and Dirac equations even though this is not apparent from the differential equations themselves.

If there is one lesson to be learned from FST, it is that if Fractals *are* present in a physical system, they can have the annoying habit of hiding underneath the resolution of differential operators. Given the central importance of the differential calculus to the language of physicists, this can account for some of the tendency of Fractals to fly under the radar of most physicists. To see them at all, you have to give up, at least temporarily, the most important tool physicists have had since the time of Newton … differentiability.

We can illustrate the odd treatment of time, and the stealth of Fractal processes in a very simple model related to isotope clocks. Consider a large number $N(t)$ of radioactive atoms in some archaeological artifact where at $t = 0$, $N(t) = N_0$ is a very large number (typically $\approx 10^{20}$). Suppose the lifetimes of all the atoms are independent identically distributed random variables from the exponential distribution with mean $\bar{T} = 1 / \lambda$ years. The expected value of $N(t)$ is then $E[N(t)] = N_0 \exp[-\lambda t]$. For very large N_0 we can ignore the discrete nature of the counting process and think of this as an idealized 'clock' that satisfies the initial value problem:

$$\frac{dN}{dt} = -\lambda N \quad N(0) = N_0. \tag{1}$$

In the isotope dating context the Archaelogist uses the inverse relation

$$t = \frac{\ln(N_0/N)}{\lambda} \tag{2}$$

to read the 'clock' for the time since the artifact was assembled. Note that as an idealization, (2) allows one to date the object for arbitrarily large values of t. In practice the clock 'breaks' from the point of view of the Archaeologist when $N(t)$ is so small compared to N_0 that (2) becomes numerically unstable with respect to fluctuations in the measured value of $N(t)$. For example in radiocarbon dating $1/\lambda = 8033$ yr. but the 'clock' is only usable up to about 60,000 years.

It has been said that the classical world is a world of objects whereas the quantum world is a world of processes. We can think of classical objects as being represented by isotope clocks in the sense that all macroscopic objects from lumps of radioactive atoms, to life forms, to galaxies, have finite lifetimes and essentially exist until their intrinsic clocks run out.

On scales far below those of our senses quantum 'objects', like electrons, evidently run forever if left alone so their 'clocks' cannot run out! To circumvent the effective lifetime of processes based on an analog of exponential decay we have to construct a means by which a stochastic process can 'forget' the initial condition that starts the clock. Here is a way of doing this that illustrates a connection to Fractal time that characteristically disappears in a continuum limit [3, 4, 5, 6].

Consider a single-loop sample path illustrated in Figure [1]. By single-loop we mean that the path starts at the origin, takes a zig-zag course out to some arbitrarily large value of t and then returns to the origin. The path crosses itself many times in the process but returns to the origin once. This will be a single 'sample-path' and an ensemble average is generated by repeating the process many times.

Notice that the particle path, on first return to the origin, constructs a chain of oriented $\xi - t$ areas. The areas and the orientations can be deduced by the 'enumerative path' sketched in Fig. [1 B]. This is the right-hand boundary of the chain of areas. We have distinguished the edges in the path according to the direction of traversal, solid for $+t$, dashed for $-t$. We can keep track of the orientation along the chain of areas by recording the state of the enumerative path. There are four states to such a path according to the four directions, $(1,1),(-1,1),(-1,-1),(1,-1)$ in the (ξ,t) plane. We will refer to these states as one to four respectively, noting that the links in an enumerative path cycle sequentially through these states like a digital clock. Despite the digital signal for each single-loop path, we can anticipate that if we allow the stochastic process to repeat many times, the expected value of the orientation will vary with t and possibly converge to some smooth function. Let us see how this happens.

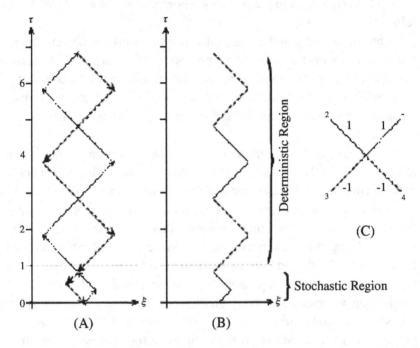

Fig. 1. (A) A single loop sample path starts on the ξ-axis, proceeds out to $t > 6$ (black path) and returns (dashed path) to the origin. In the 'interaction' region $t < 1$ the places where the path changes direction or crosses the return path are determined by a stochastic process. Once the path has escaped the interaction region the path is deterministic and forms a chain of equal areas from the last corner/crossing point in the interaction region. The orientation of successive areas alternates sign. (B) The right enumerative path concatenates the right hand boundary of the chain of oriented areas. The alternating colouring indicates the contribution to the sign of the orientated area. (C) The four states correspond to edges in the enumerative path. The colour indicates the sign of the oriented area to the left of the path.

Let $U_P(t)$ be a four component random variable that indicates the state of a particle's enumerative path at time t.

$$U_P(t) \in \left\{ \begin{pmatrix} 1 \\ 0 \\ 0 \\ 0 \end{pmatrix}, \begin{pmatrix} 0 \\ 1 \\ 0 \\ 0 \end{pmatrix}, \begin{pmatrix} 0 \\ 0 \\ 1 \\ 0 \end{pmatrix}, \begin{pmatrix} 0 \\ 0 \\ 0 \\ 1 \end{pmatrix} \right\} \tag{3}$$

Each event in the interaction region changes $U_P(t)$ by shifting it to the next element in the list, where the list is just a repetition of the four elements in (3). Thus each event corresponds to an application of the shift matrix:

$$S = \begin{pmatrix} 0 & 0 & 0 & 1 \\ 1 & 0 & 0 & 0 \\ 0 & 1 & 0 & 0 \\ 0 & 0 & 1 & 0 \end{pmatrix}.$$

Given that the waiting times between events are exponential and independent with rate λ, the number of events in time t will be Poisson-distributed so that if the paths all start with state distribution $U_P(0)$ the expected state occupation vector is

$$E[U_P(t)] = \sum_{k=0}^{\infty} \frac{(\lambda t)^k S^k}{k!} \exp(-\lambda t) U_P(0) \tag{4}$$

for $t < t_0$. $E[U_P]$ gives us the expected population of each state at time t. Each component of $E[U_P]$ will be non-negative and the four-component vector is normalized so the vector itself is a probability mass function(PMF) giving the expected proportion of paths in each state. Calculating (4) is straightforward, but to see why the result is simple we can make a change of variables from the probability vector $U_P(t)$ to new variables $\phi(t)$ in which the first two components are even sums of probabilities and the last two components are odd *differences* of probabilities. With the change of variables the shift matrix is transformed to:

$$S_{E/O} = \begin{pmatrix} 0 & 1 & 0 & 0 \\ 1 & 0 & 0 & 0 \\ 0 & 0 & 0 & -1 \\ 0 & 0 & 1 & 0 \end{pmatrix} \tag{5}$$

The transformed shift matrix operates separately on the even and odd components of ϕ and the resulting expected signal, up to a multiplicative constant is:

$$E[\phi(t)] = \sum_{k=0}^{\infty} \frac{(\lambda t)^k S_{E/O}^k}{k!} \phi(0)$$

$$= \begin{pmatrix} \cosh(\lambda t) & \sinh(\lambda t) & 0 & 0 \\ \sinh(\lambda t) & \cosh(\lambda t) & 0 & 0 \\ 0 & 0 & \cos(\lambda t) & -\sin(\lambda t) \\ 0 & 0 & \sin(\lambda t) & \cos(\lambda t) \end{pmatrix} \phi(0) \tag{6}$$

The even/odd block structure persists through the ensemble average and we can consider the two systems separately *after the ensemble average*. The upper even block displays the exponential decay to equilibrium typical of terrestrial clocks. If we allowed an arbitrarily long interaction time $t_0 \gg 1$ the upper two states would approach their equilibrium configuration of equal probability and would represent a form of the isotope clock, satisfying the familiar equation (1). The counting process we rely on to construct probabilities naturally falls into the 'even' eigenspace corresponding to the first two components of ϕ.

The lower odd block however returns an oscillation. For appropriate choice of λ the lower two components satisfy the equation:

$$\frac{d\phi}{dt} = i\lambda\sigma_y\phi. \tag{7}$$

This equation actually holds for *for all* $t > 0$. Here σ_y is the second Pauli matrix. This is the same as the isotope clock equation (1) except we have replaced $-\lambda$ by $i\lambda\sigma_y$. Instead of the decaying exponential solutions of (1) the solutions of (7) are the familiar trigonometric functions. These oscillate at a fixed frequency and as clocks, never run out. From an 'Archaeological' point of view our stochastic process is keeping track of time as the argument of a complex exponential!

Notice that the effect of looking at the odd part of this simple stochastic process is that it behaves as if the t variable of the even process has been replaced by it ($i\sigma_y$ is a real, 2×2 representation of i). It appears to be this replacement that gives the odd metric of Minkowski space ... development of this model shows that the σ_y in (7) becomes the β matrix in the Dirac equation [7]. It is also this (Fractal induced) analytic continuation that gives us the transition from the probability density functions of classical probability to the 'wavefunctions' of quantum mechanics [6]. The Fractal is of course lurking in the scattering of the process in the region $0 < t < t_0$.

The Fractal nature of time, the underlying statistical mechanics and the connection to ephemeral isotope clocks in a world of classical objects are all effectively hidden underneath the limit implied by taking an expectation over all sample paths. Odd indeed!

References

1. G. N. Ord. Fractal space-time a geometric analog of relativistic quantum mechanics. *J. Phys. A*, 16:1869–1884, 1983.
2. B. B. Mandelbrot. *Fractals: Form, Chance, and Dimension*. San Francisco: Freeman, 1977.
3. G. N. Ord and J. A. Gualtieri. The Feynman propagator from a single path. *Phys. Rev. Lett*, 89(25):250403–250407, 2002.
4. G. N. Ord and R. B. Mann. Entwined paths, difference equations and the Dirac equation. *Phys. Rev. A*, 67, 2003.
5. G. N. Ord, R. B. Mann, E. Harley, Zenon Harley, Qin Qin Lin, and Andrew Lauritzen. Numerical experiments in relativistic phase generation through time reversal. *Advanced Studies in Theoretical Physics.*, 3(3):99–130, 2009.
6. G. N. Ord. Quantum mechanics in a two-dimensional spacetime: What is a wavefunction? *Annals of Physics*, 324(6):1211–1218, June 2009.
7. G. N. Ord and R. B. Mann. Entwined paths, difference equations and the Dirac equation. *Phys. Rev. A*, 67(2):022105, Feb 2003.

The Theory of Scale Relativity and Fractal Space-Time

Laurent Nottale

Directeur de Recherche CNRS
LUTH, Observatoire de Paris-Meudon, France

The theory of scale relativity is a geometric framework constructed in terms of a fractal and nondifferentiable continuous space-time. This theory leads (i) to a generalization of possible physically relevant fractal laws, written as a partial differential equation acting in the space of scales, and (ii) to a new geometric foundation of quantum mechanics and gauge field theories and their possible generalizations.

One of the main concerns of scale relativity is about the foundation of quantum mechanics. As is now well known, the principle of relativity (of motion) underlies the foundation of most of classical physics. Now, quantum mechanics, though it is harmoniously combined with special relativity in the framework of relativistic quantum mechanics and quantum field theories, seems, up to now, to be founded on different grounds. Actually, its present foundation in standard quantum theory is mainly axiomatic, i.e., it is based on postulates and rules which are not derived from any underlying, more fundamental principle.

The theory of scale relativity suggests an original solution to this fundamental problem. Namely, in its framework, quantum mechanics may indeed be founded on the principle of relativity itself, provided this principle (which, up to now, has been applied to position, orientation and motion) be extended to scales. The definition of reference systems can be generalized by including variables characterizing their scale. Then the possible transformations of these reference systems can be generalized by adding to the relative transformations already accounted for in standard theories of relativity (translation, velocity and acceleration of the origin, rotation of the axes) the transformations of these scale variables – namely, their relative dilations and contractions. In the framework of such a newly generalized relativity theory, the laws of physics may be given a general form that transcends and includes both the classical and the quantum laws, allowing us, in particular, to study in a renewed way the poorly understood nature of the classical to quantum transition.

A related important concern of the theory is the question of the geometry of space-time at all scales. In analogy with Einstein's construction of general relativity of motion, which is based on the generalization of flat space-times to curved Riemannian geometry, it is suggested, in the framework of scale relativity, that a new generalization of the description of space-time is now needed, towards a still continuous but now nondifferentiable and fractal geometry (i.e., a geometry in which the various quantities are explicitly dependent on the scale of observation or measurement). This, therefore, requires the development of new mathematical and physical tools in order to implement such a generalized description, which goes far beyond the standard view of differentiable manifolds.

In this framework, the equations of motion in such a fractal space-time are to be written as geodesic equations (i.e., equations which express the optimization of the proper time), under the constraint of the principles of scale relativity and of scale covariance. To this purpose, covariant derivatives are constructed that implement the various effects of the nondifferentiable and fractal geometry.

As a first theoretical step, the laws of scale transformation that describe the new dependence on resolutions of physical quantities are obtained as solutions of differential equations acting in the space of scales. This leads to several possible levels of description for these laws, from the simplest scale-invariant laws to generalized laws with variable fractal dimensions. These include log-periodic laws and log-Lorentz laws of "special scale-relativity", in which the Planck scale is identified with a minimal, unreachable scale, invariant under scale transformations (in analogy with the special relativity of motion in which the velocity of light in vacuum c is invariant under motion transformations).

The second theoretical step amounts to a description of the effects induced by the fractal structures of geodesics on motion in standard space (of positions and instants). Their main consequence is the transformation of classical dynamics into a generalized, quantum-type dynamics.

The theory makes it possible to define and derive from relativistic first principles both the mathematical and physical quantum tools (complex wave functions, then spinor, bispinor and multiplet wave functions) and the equations of which these wave functions are solutions: a Schrödinger-type equation (more generally a Pauli equation for spinors) is derived as an integral of the geodesic equation in a fractal space, then the Klein-Gordon and Dirac equations in the case of a full fractal space-time. Finally, gauge fields and gauge charges can also be constructed from a geometric re-interpretation of gauge transformations as being nothing else but scale transformations of the very small scale structures of the geodesics in a fractal space-time (by which the "particles" are identified).

This theory has been applied to various sciences, and in several cases, its theoretical predictions have been validated by new or updated observational and experimental data. This includes applications in physics and cosmology (value of the QCD coupling and of the cosmological constant), to astrophysics and gravitational structure formation (distances of extra-solar planets to their stars, of Kuiper belt objects, value of solar and solar-like star cycles, structures over many scales, from the planetary to the extragalactic scales), to life sciences (log-periodic law for species punctuated evolution, human development and society evolution), to Earth sciences (log-periodic deceleration of the rate of California earthquakes and of various earthquake after-shocks, critical law for the extent of Arctic sea ice) and tentative applications to systems biology (processes of morphogenesis, emergence of prokaryotic and eukaryotic cellular structures, cell confinement, duplication and branching).

Selected Bibliographic References

L. Nottale, Fractal Space-Time and Microphysics: Towards a Theory of Scale Relativity. World Scientific, 1993, 333 pp.

M.N. Célérier and Nottale L., Quantum-classical transition in scale relativity, in: J. Phys. A37 (2004), 931, http://arXiv.org/abs/quant-ph/0609161

L. Nottale, Célérier M.N. & Lehner T., Non-Abelian gauge field theory in scale relativity, in: J. Math. Phys. 47 (2006), 032303, http://arXiv.org/abs/hep-th/0605280

L. Nottale and Célérier M.N., Derivation of the postulates of quantum mechanics from the first principles of scale relativity, in: J. Phys. A: Math. Theor. 40, (2007) pp. 14471-14498, http://arXiv.org/abs/0711.2418 [quant-ph].

C. Auffray and Nottale L., Scale relativity theory and integrative systems biology. 1. Founding principles and scale laws, in: Progress in Biophysics and Molecular Biology, 97 (2008) pp. 79-114 http://www.luth.obspm.fr/~luthier/nottale/ arPBMB08AN.pdf

L. Nottale and Auffray C., Scale relativity theory and integrative systems biology. 2. Macroscopic quantum-type mechanics, in: Progress in Biophysics and Molecular Biology, 97, (2008) pp. 115-157, http://www.luth.obspm.fr/~luthier/nottale/ arPBMB08NA.pdf

L. Nottale and P. Timar, Relativity of Scales: Application to an Endo-Perspective of Temporal Structures, in: S. Vrobel, O.E. Rössler, T. Marks-Tarlow (Eds.), Simultaneity – Temporal Structures and Observer Perspectives, World Scientific, Singapore 2008.

L. Nottale, The Theory of Scale Relativity. Fractal Space-Time, Nondifferentiable Geometry and Quantum Mechanics. Imperial College Press, 596 pp. 2010.

Appendix E

A Very Concise Summary of the Main Ideas of E-Infinity Cantorian Spacetime Theory

Mohamed El Naschie

There are three versions of fractal spacetime theory. The first is that of Garnet Ord which by and large tries to relate quantum mechanics to a specific realistic model in the spirit of what Einstein did for diffusion. The second version is due to Laurent Nottale which he himself describes as scale relativity theory where the speed of light is replaced by the Planck length as a cut off quantity. Both versions imply fractality of spacetime itself. The third version which we would like to summarize here in a nutshell derives from the outset the geometry of spacetime and finds that it is not just any kind of fractal but a specific type of fractal. It is an infinite dimensional random Cantor set and consequently it is based on the simplest and most natural of all possible fractals.

Let us begin at the beginning. The reader will recall that the simplest kind of fractal is the triadic Cantor set which lives in one dimension. Take the middle third from the unit interval but leave behind the n point and go on doing that ad infinitum and you will end with an infinite number of points constituting the triadic Cantor sets. Since you have removed the entire length of the unit interval, whatever is left has no length. We say it is measure zero. By contrast it has substantial dimension, namely $\ln 2 \div \ln 3$ whichis about 0.630929. This is the so called Hausdorff or fractal dimension. On the other hand if this set is constructed randomly in the way described in many E-infinity publications then something quite remarkable comes out as a Hausdorff dimension, namely the golden mean which is equal to 0.618033. In E-infinity theory it turns out that we can model quantum spacetime geometry by an infinite dimensional hierarchal random Cantor set. There is a specific way to determine the expectation value of the Hausdorff dimension of such a space and it turns out to be equal to the inverse of the golden mean to the power of 3. In other words, the Hausdorff dimension of such Cantorian spacetime is equal to 4.236067977. This is exactly equal to 4 plus the golden mean to the power of 3. One of the most mathematical aspects of E-infinity Cantorian spacetime is the link which it established between the fractal nature of the geometry and a certain type of topological dimension which is well known in the theory of transfinite dimensions. This dimension is called the Menger-Urysohn dimension which is a sophisticated and very precise way of describing the topological dimension of a structure. E-infinity links the Hausdorff dimension with the Menger-Urysohn dimension and at the end we have really three dimensions which together and only together can give a precise description of quantum spacetime. It turns out that the topological Menger-Urysohn dimension is exactly 4 in agreement with our observation at low energy resolution. The Hausdorff dimension on the other hand is 4.236067977. Never the less the formal dimension is infinitely large because our Cantorian space is made up of an infinite number of elementary Cantor sets with progressively smaller and smaller Hausdorff dimensions. It follows from the preceding discussion that the Cantorian spacetime proposal implies that our space is not only resolution dependent with regard to the fine structure but even the number of dimensions which we could observe is resolution dependent.

Another crucial point which was used to resolve many paradoxes inherent in conventional quantum mechanics, particularly the two-slit experiment with quantum particles could be solved with unheard of simplicity when we note that the Menger-Urysohn dimension of a Cantor set with a Hausdorff dimension equal to the golden mean is zero while the Menger-Urysohn dimension of a Cantor set with a Hausdorff dimension equal to the golden mean squared is equal to minus 1. Thus we have in a natural way negative dimensions which follows from a general and powerful mathematical theory making the need for anti-commuting Grassmannian coordinates redundant.

There is a very simple model for E-infinity theory in two dimensions. This model is nothing else but the familiar Penrose fractal tiling. On the other hand Penrose fractal tiling was recognized some time ago by Alain Connesto be a realization of non-commutative geometry. In other words E-infinity theory is yet another version and probably a more general form of non-commutative geometry. Based on these ideas El Naschie and his group, particularly Ji-Huan He in China, L. Marek-Crnjac in Slovenia, G. Iovane in Italy were able to link the theory with the holographic principles as well as with the exceptional Lie symmetry groups in particular E8.

Using the preceding theory one is able to derive the mass spectrum of elementary particles of the standard model as well as all the constants of nature including the fine structure constant as well as Newton's gravity constant.

For a historical review of the theory the reader is directed to a very readable paper by L. Marek-Crnjac entitled A Short History of Fractal Spacetime. In addition to this essential reading two review articles by El Naschie offer a simple introduction to E-infinity theory. The first paper which we recommend is entitled A Review of E-Infinity and The Mass Spectrum of High Energy Particle Physics published in *Chaos, Solitons & Fractals*, 19, p. 209-236 (2004). The second and more up to date paper is The Theory of Cantorian spacetime and High Energy Particle Physics (An Informal Review) published in *Chaos, Solitons & Fractals*, 41(5), p. 2635-2646 (2009).

To sum up while at the low energy resolution of classical physics, spacetime appears to be smooth and Euclidean and while it is non-Euclidean and curved at the astronomical distance of general relativity, it is discontinuous, random and chaotic at the resolution of quantum mechanics and high energy physics. It all depends upon the resolution with which we probe spacetime. The link between the fractal Cantorian randomness of micro spacetime and particle physics arises from the same basic philosophy which Einstein used to attribute gravity to the curvature of spacetime.

Recently, experimental verification of E8 symmetry has been provided by researchers of the Helmhotz-Zentrum in Berlin, in co-operation with colleagues from Oxford and Bristol Universities and the Rutherford Appleton Laboratory. The signatures of the nanoscale symmetry they measured showed the same attributes as the golden ratio. The results were published in R. Coldea, D.A. Tennant, E.M. Wheeler, E. Wawrzynska, D. Prahhakaran, M. Telling, K. Habicht, P. Smeibidl, K. Kiefer: Quantum Criticality in an Ising Chain: Experimental Evidence for Emergent E8 Symmetry, in: *Science*, Vol. 327 (8 Jan. 2010), No. 5962, pp. 177-180.

Index